Jürgen O. Besenhard,
Werner Sitte, Franz Stelzer,
Heinz Gamsjäger (eds.)

Electroactive Materials

SpringerWienNewYork

Prof. Dr. Jürgen O. Besenhard
Institute of Chemical Technology of Inorganic Materials, Graz University of Technology,
Graz, Austria

Prof. Dr.Werner Sitte
Institute of Physical Chemistry, University of Leoben,
Leoben, Austria

Prof. Dr. Franz Stelzer
Institute of Chemistry and Technology of Organic Materials, Graz University of Technology,
Graz, Austria

Prof. Dr. Heinz Gamsjäger
Institute of Physical Chemistry, University of Leoben,
Leoben, Austria

Typesetting: Thomson Press (India) Ltd., New Delhi
Printing: Manz Crossmedia, A-1051 Wien
Binding: Papyrus, A-1100 Wien
Printed on acid-free and chlorine-free bleached paper
SPIN: 10833609

With 67 Figures

CIP data applied for

Special Edition of
Monatshefte für Chemie/Chemical Monthly Vol. 132, No. 4, 2001

ISBN 978-3-211-83655-2 Springer-Verlag Wien New York

Editorial

Electroactive materials are playing an ever increasing role in science and technology. At present, the wide range of applications for these materials includes electrodes and membranes for electrochemical energy conversion and storage, electroceramic devices and sensors, organic diodes, magnetic and optical devices, and photoresists. To understand the complex functionality of the materials involved in various devices necessarily requires a long-term, interdisciplinary fundamental research effort by chemists, physicists, materials scientists, and engineers.

To support intensive research in this field, the special research program (SFB) 'Electroactive Materials' was established by the Austrian Science Fund in 1996, integrating already existing research activities at the Graz University of Technology (*Erzherzog-Johann-Universität*) and the University of Graz (*Karl-Franzens-Universität*). The major goal of this special research program is the preparation, characterization, and application of technologically relevant organic and inorganic materials with special electrochemical, electric, dielectric, and optoelectronic properties. The individual projects cover research activities on organic as well as inorganic electroactive materials for new types of devices and the improvement of the functional materials themselves.

With this special issue 'Electroactive Materials' the Editors of *Chemical Monthly* offered the SFB a forum for presenting an overview of the current research activities. These involve the synthesis of electroactive organic and inorganic materials as well as new methods for their characterization and application. We are also very pleased to include an Invited Review by *Serdar Sariciftci* on recent developments in conjugated polymer-based plastic solar cells. Thus, the contributed articles cover not only a broad spectrum of aspects in synthesis and characterization of organic materials with properties qualifying them for use in various opto-electronic and electro-optic devices, but also recent results of research on materials for applications in advanced devices for energy conversion and storage. Each article highlights an actual topic of the respective research group, often including a broader introduction for readers not fully involved in the field.

We are proud to be able to offer 12 articles dealing with purely organic materials such as conjugated polymers and organosilanes, thin-film zinc/manganese dioxide electrodes, studies on the anode/electrolyte interface in lithium ion batteries, a novel technique for manufacturing highly conductive composite materials by dip-coating insulators with conductive materials, a new method for conductivity relaxation measurements on mixed conductors using the *van der Pauw* geometry, the application of surface science to the investigation of thin films and interfaces of electroactive organic materials, preparation and radical oligomerization of an Fe(II) complex without loss of spin-crossover properties, and a paper dealing with phase gratings in photoreactive polymers as a way to optically pumped organic lasers. Last but not least, one contribution reports on important progress in high-spatial resolution elemental analysis and mapping by energy-filtering transmission electron microscopy and electron energy-loss spectrometry. All contributors hope that their individual papers could also make visible new trends in the various research fields of electroactive materials.

We want to thank the Austrian Science Fund, the Austrian National Bank (OENB), the Federal State of Styria, and the City of Graz as well as the Graz University of Technology and the University of Graz for financial support. In addition, we gratefully appreciate the invaluable help of *H. Kalchhauser* in publishing this special issue.

Jürgen O. Besenhard
Werner Sitte
Franz Stelzer
Heinz Gamsjäger
Issue Editors

Contents

Invited Review

Recent Developments in Conjugated Polymer Based Plastic Solar Cells

Christoph J. Brabec and **Serdar N. Sariciftci***

Linzer Institut für Organische Solarzellen, Abteilung für Physikalische Chemie, Johannes Kepler Universität Linz, A-4040 Linz, Austria

Summary. Recent developments on photovoltaic elements based on solid state composites of conjugated, semiconducting polymers mixed with buckminsterfullerene are reviewed. The photoinduced charge transfer from donor-type semiconducting conjugated polymers onto acceptor-type conjugated polymers or acceptor molecules such as buckminsterfullerene is reversible, ultrafast (within 100 fs) with a quantum efficiency approaching unity, and the charge separated state is metastable (up to ms at 80 K). This phenomenon of photoinduced electron transfer leads to a number of potentially interesting applications which include, among others, sensitization of the photoconductivity, reverse saturable absorption (optical limiting), and photovoltaic phenomena. Recent studies on the realization of photovoltaic elements with 3% power conversion efficiency are reported.

Keywords. Organic photovoltaics; Solar cells; Conjugated polymers; Fullerenes.

Introduction

Recent developments in realizing light emitting diodes based on conjugated semiconducting polymers (Fig. 1) have stirred a large scientific and technological effort [1–5]. If photoinduced free charge carrier generation is allowed at the same time, the electroluminescence device shows light emission under forward bias and a significant photocurrent under a reverse bias field (dual function) [6]. Using the devices for photodetection under reverse bias, the potential difference between the electrodes has to be high enough to overcome the *Coulomb* attraction of the photogenerated excitons. Otherwise, the absorbed photons will mainly create excitons which decay geminately (either radiatively, with photoluminescence, or non-radiatively). Some improvement has been reported for photodiodes utilizing a *Schottky* type junction formed between the conjugated polymer and one of the metal electrodes; however, the problem of inefficient charge generation in conjugated polymers has not been resolved by this approach [7–9]. To overcome

* Corresponding author

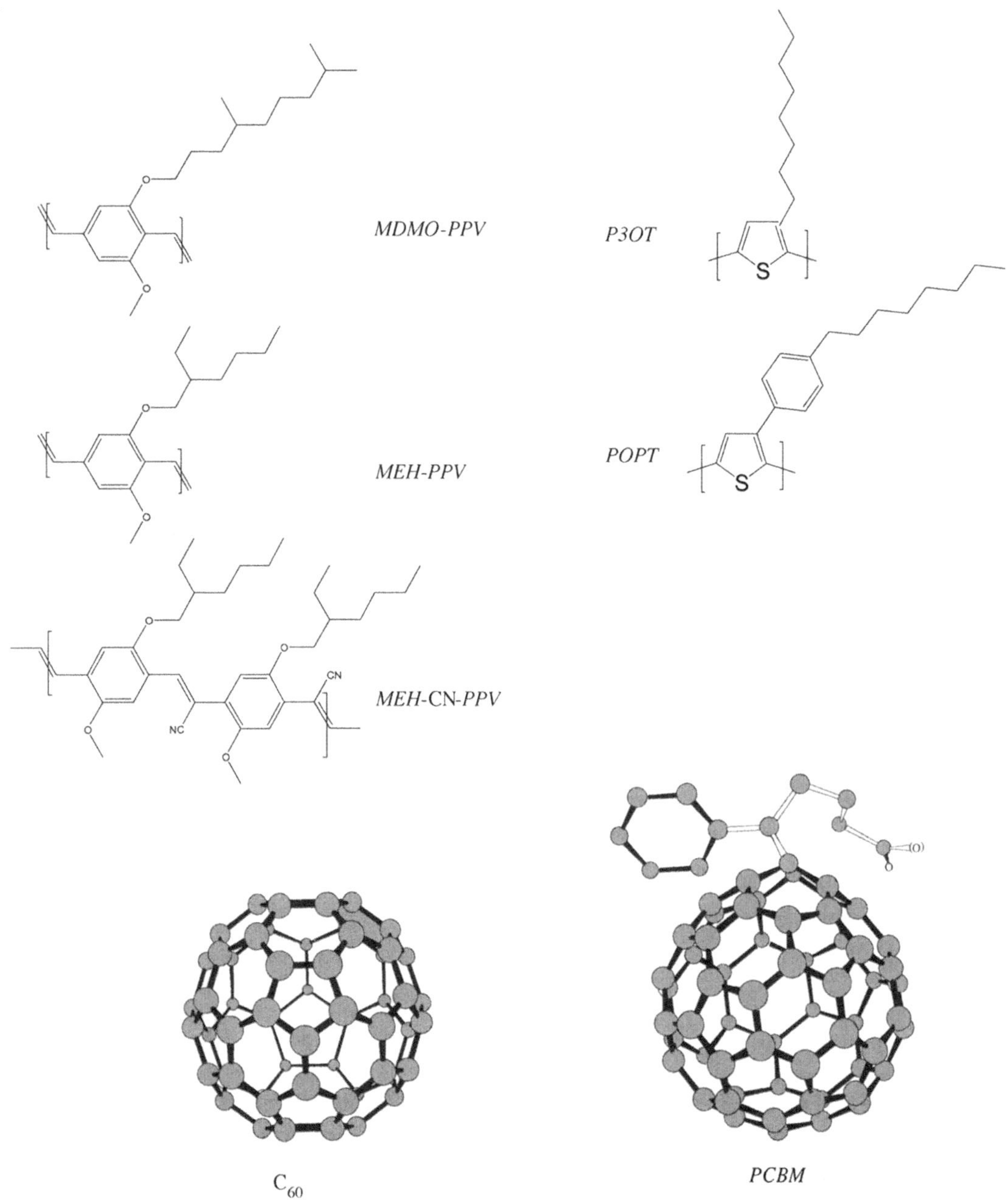

this limitation of photoinduced charge carrier generation, a donor/acceptor (dual molecule) approach has been suggested [10–13] consisting of a composite thin film with a conjugated polymer/fullerene mixture. In such single composite photoactive films, a bulk heterojunction is formed between the electron donors and acceptors (Fig. 2) [11, 12], and the efficiency of photogeneration of charges is near 100%.

Many conjugated polymers (see Fig. 1) in their undoped, semiconducting state are electron donors upon photoexcitation (electrons promoted to the antibonding π^* band). The idea of using this property in conjunction with a molecular electron acceptor to achieve long-living charge separation was based on the stability of the

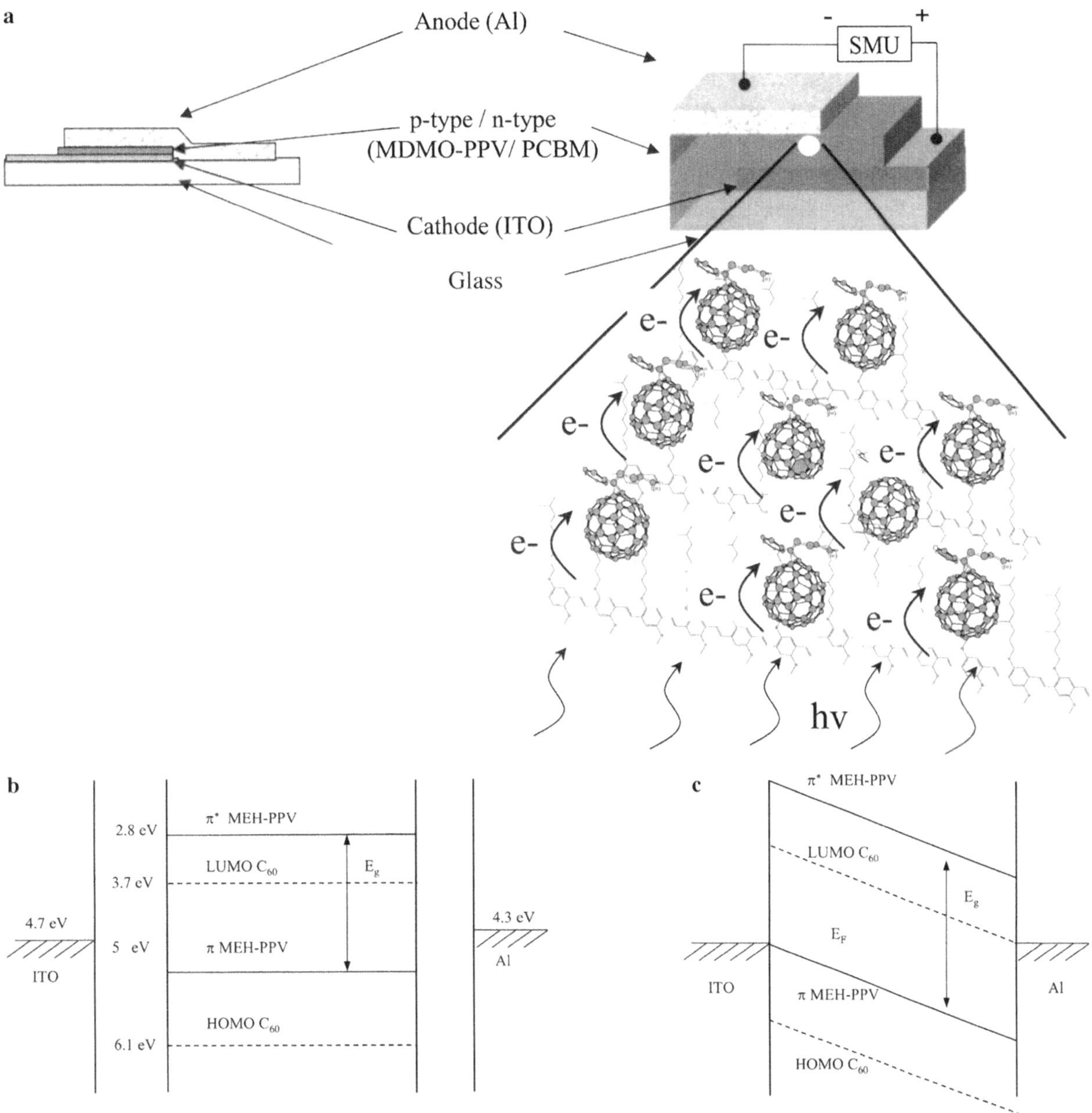

composite formed from the interpenetrating donor/acceptor network plotted with the device structure for such a kind of junction (a); the diagrams with the energy levels of a *MDMO-PPV*/C_{60} bulk heterojunction system (as an example) under flat band conditions (b) and under short-circuit conditions (c) do not take into account possible interfacial layers at the metal/semiconductor interface

photoinduced nonlinear excitations (such as polarons) on the conjugated polymer backbone. Once the photoexcited electron is transferred to an acceptor unit, the resulting cation radical (positive polaron) species on the conjugated polymer backbone is known to be highly delocalized, mobile, and stable as shown in electrochemical and/or chemical oxidative doping studies. Analogous to the chemical doping process, we will term the photoinduced electron transfer from the conjugated polymer donor onto an acceptor moiety as photodoping.

Background of Conjugated Polymer Based Solar Cells

Independently, *Sariciftci et al.* and *Yoshino et al.* have reported studies on the photophysics of mixtures of conjugated polymers with fullerenes [10, 14–22, 25]. The observations clearly evidenced an ultrafast, reversible, metastable photoinduced electron transfer from conjugated polymers onto buckminsterfullerene in solid films; for a schematic description of this phenomenon, see Fig. 2. Using this molecular effect at the interface of bilayers consisting of a semiconducting polymer (poly-(2-methoxy-5-(2′-ethyl-hexoxy)-*p*-phenylene) vinylene, hereafter referred to as *MEH-PPV*) and C_{60} films, diodes with rectification ratios in the order of 10^4 and a photovoltaic effect [11, 23] have been manufactured. Significant improvement of the relatively low charge collection efficiency of the D/A bilayer was achieved by using phase separated composite materials, processed through control of the morphology of the phase separation into an interpenetrating network (bulk heterojunction). The power conversion efficiency of solar cells made from *MEH-PPV*/ methanofullerene composites was subsequently increased dramatically [12]. Parallel, the group of *Alan Heeger* in Santa Barbara and *Richard Friend* in Cambridge, developed an approach using acceptor-type conjugated polymers in an interpenetrating polymer-polymer composite with *MEH-PPV*, yielding polymeric photovoltaic devices with efficiencies comparable to fullerene mixed devices [13, 24, 25].

Using sub-picosecond photoinduced absorption studies [15, 26, 27], ultrafast (< 1 ps) photoinduced electron transfer has been demonstrated in the composites of conjugated polymers with fullerenes. Very recently, the forward transfer of the photoexcited electrons from conjugated polymer donors onto *PCBM* (cf. Fig. 1) acceptors was resolved by pump-probe experiments with an unprecedented time

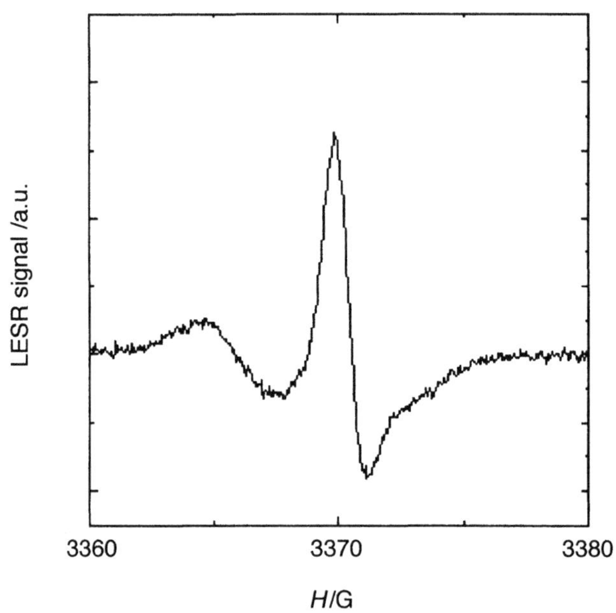

Fig. 3. Integrated light induced electron spin resonance spectra of *MDMO-PPV/PCBM* upon successive illumination with a 2.41 eV argon ion laser (reproduced with permission of the American Physical Society after Ref. [31])

resolution of 10 fs [28]. In these studies, the relaxation of the photoinduced excitations on the polymer chain switched from radiative (intense stimulated emission in pristine polymer, $\Delta T > 0$) to non-radiative (electron transfer in composites, photoinduced absorption, $\Delta T < 0$) upon mixing *PCBM* into the polymer matrix. This experimental setup allowed to time-resolve the forward photoinduced electron transfer time with ~40 fs after an initial relaxation (*Kasha* relaxation). As such, this photoinduced electron transfer is indeed ultrafast, resulting in a quantum yield of photoinduced charge generation of 100%.

The time-resolved transient photocurrent (PC) of *MEH-PPV*/C_{60} composites with different C_{60} content [29] shows an increase of initial photocurrent by an order of magnitude upon admixture of 1% of C_{60} [29]. Figure 3 shows the integrated ESR signal upon illuminating the *MDMO-PPV*/*PCBM* (cf. Fig. 1) composites with light of $h\nu = E_{\pi-\pi^*}$, where $E_{\pi-\pi^*}$ is the energy gap of the conjugated polymer (donor). Two photoinduced ESR signals can be resolved, one at $g = 2.0026$ and the other at $g = 1.9997$. The higher g-value line is assigned to the conjugated polymer cation (polaron), and the lower g-value line to the $PCBM^-$ anion [30, 31].

Other Organic Solar Cell Architectures

Organic bilayers for photovoltaic applications have been investigated heavily during the last couple of decades (for a summary, see for example Refs. [11, 32–35]). *Tang* demonstrated [33] in his pioneering work using a vacuum evaporated copper-phthalocyanine (250 Å)/perylene derivative (450 Å) system sandwiched between In_2O_3 and silver electrodes a white light photovoltaic power conversion efficiency of around 1%. Other examples in the literature exist on the fabrication of solar cells based on small molecular dyes as well as donor-acceptor systems (see for example Refs. [36, 37] and references therein). Using electron transfer from organic dyes onto inorganic nanoparticles, charge separation and photovoltaic conversion with *ca.* 10% efficiency have been demonstrated (see for example Refs. [38, 39] and references therein). *Yamashita* and coworkers have reported a bilayer photodiode based on the organic donor tetrathiafulvalene (*TTF*) and C_{60} [40]. *Meissner et al.* [41–44] enhanced the performance of a zinc-phthalocyanine (Zn*Pc*)/N,N′-dimethyl-perylene-3,4,9,10-tetracarboxylic diimide (*MPP*) bilayer device by inserting a 30 nm thick C60-Zn*Pc* composite layer between Zn*Pc* (50 nm) and *MPP* (20 nm) [42]. Last but not least, in photovoltaic cells made of bilayers of conjugated polymer and C_{60}, monochromatic energy efficiencies as high as 1% and IPCE efficiencies as high as 15% have been measured [45, 46].

The Bulk Heterojunction Concept

For photovoltaic cells made from pure conjugated polymers, energy conversion efficiencies were typically 10^{-3}–10^{-1}%, too low to be used in practical applications [47–49]. Consequently, interpenetrating phase separated D/A network composites, *i.e.* bulk heterojunction, would appear to be ideal for photovoltaic materials [12]. Through control of the morphology of the phase separation into an interpenetrating network, one can achieve a high interfacial area within a bulk material. If any point in the composite is within a few nanometers of a D/A interface, such a composite is a called a bulk D/A heterojunction material. If the D/A networks in a device are bicontinuous, as shown schematically in Fig. 2, the collection efficiency can be equally efficient.

It is important to note that the idealized schematic description in Fig. 2 is by far not realistic in any actual composite sample studied here. As shown below using AFM pictures, it is rather a disordered juxtaposition of islands of its two components. In addition, it is well known that polymer blends frequently tend to cleave into separate phases rather than to form one uniform medium. Decreasing the size of the islands results directly in an increase of the effective contact area between both components.

Important progress has been made towards creating bulk D/A heterojunction materials [12, 13, 24, 50]. *Gang Yu et al.* utilized the bulk heterojunction formed in a solid state composite of conjugated polymers and methanofullerene *PCBM* [12] with a short-circuit current of $I_{SC} = 0.5\,\mathrm{mA/cm^2}$ under $20\,\mathrm{mW/cm^2}$ illumination, approximately two orders of magnitude higher than that of pure *MEH-PPV* tunnel diodes as well as of the *MEH-PPV*/C_{60} heterojunction device described in the previous section. Uniform films with even higher concentrations of *PCBM* were cast from 1,2-dichlorobenzene solutions containing weight ratios of *MEH-PPV:PCBM* up to 1:4.

The groups of *Richard Friend* in Cambridge and *Alan Heeger* in Santa Barbara also simultaneously constructed a photovoltaic bulk heterojunction using two different conjugated polymers [24, 13, 50]. Both approaches employ *MEH-PPV* as donor conjugated polymer in composite with a cyano substituted *MEH-PPV*-type polymer. The cyano substitution was originally introduced into the light emitting device materials because of the lowered LUMO level of this derivative relative to the unsubstituted polymer [51]. This lower LUMO of the high electron affinity material eases the electron injection from a medium workfunction metal such as aluminum. Recent reports on polymer/polymer devices led to considerable improvement of the general efficiencies of these types of photovoltaic cells [50] using different stochiometry of two laminating layers creating a gradient for hole conducting and electron conducting components in the bulk heterojunction.

At the University of Linz, large area ($6\,\mathrm{cm}\times 6\,\mathrm{cm}$ and, in a later stage, $10\,\mathrm{cm}\times 15\,\mathrm{cm}$) photovoltaic elements on flexible *ITO* (indium tin oxide) coated plastic (*PET*) substrates have been prepared. A picture of such an element is shown

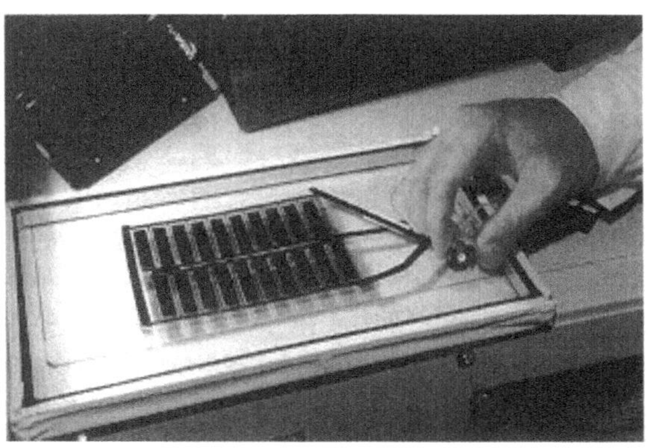

Fig. 4. Picture of a large area plastic solar cell running a small motor

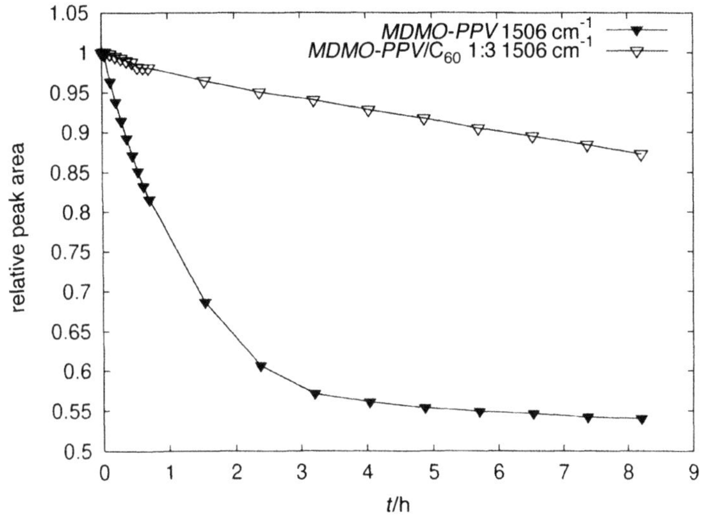

Fig. 5. Time dependence of specific absorption bands: *MDMO-PPV* at $1506\,cm^{-1}$ (downward full triangles) and 1:3 mixture of *MDMO-PPV*/C_{60} at the *MDMO-PPV* band at $1506\,cm^{-1}$ (downward open triangles) during controlled degradation test (reproduced by permission of Wiley-VCH from Ref. [35])

in Fig. 4. Large area devices with AM 1.5 simulated efficiencies around 2% can now be routinely fabricated.

Furthermore, systematic studies showed that upon addition of the fullerene derivatives to the conjugated polymer matrix, the stability of the matrix is increased [52, 53]. Figure 5 clearly shows the enhancement of the stability of the conjugated polymer/*PCBM* composite compared to the conjugated polymer component alone. Since the studies are directly related to specific infrared absorption of the conjugated polymer itself, it is concluded that the stability of the conjugated polymer component is increased upon addition of C_{60}. This is proposed to originate from the photostabilizing effect of the fullerences because of the ultrafast photoinduced electron transfer:

i) Since the electrochemical potentials of the excited conjugated polymer (one electron excited to the LUMO level) and of the excited fullerene component (one hole left in the HOMO level) are both very high, it might immediately lead to direct electrochemical interactions with ambient air humidity *etc.* The ultrafast electron transfer from this LUMO level of the conjugated polymer onto the LUMO of C_{60} (or hole transfer from the HOMO level of C_{60} onto the HOMO of the conjugated polymer) immediately empties the excited state of the polymer (or of the fullerene) and lowers the electrochemical energy by bringing it down to more stable potentials.

ii) The intersystem crossing of the polymer (or the fullerene) produces triplet excited states, and these may in turn react by energy transfer creating singlet oxygen. This highly reactive form of oxygen is expected to react with the polymer backbone, creating carbonyl type defects and thus eliminating conjugation. By the ultrafast photoinduced electron transfer, the intersystem crossing to the triplet state is also completely quenched as shown above.

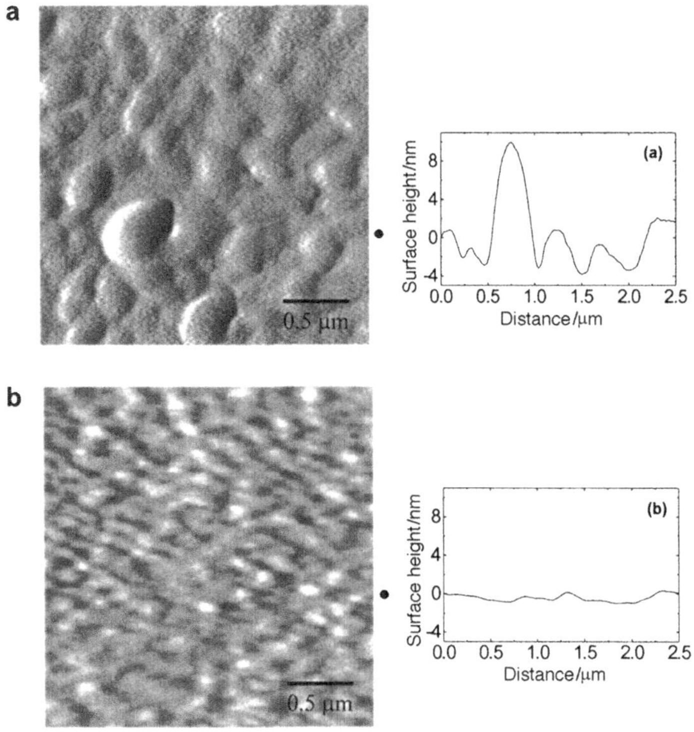

Fig. 6. AFM images showing the surface morphology of *MDMO-PPV:PCBM* (1:4 w/w) blend films with a thickness of approximately 100 nm and the corresponding cross sections; (A) spin-coated from a toluene solution, (B) spin-coated from a chlorobenzene solution; the images show the first derivative of the actual surface heights; the cross-sections of the true surface heights for the films were taken horizontally from the points indicated by the dots

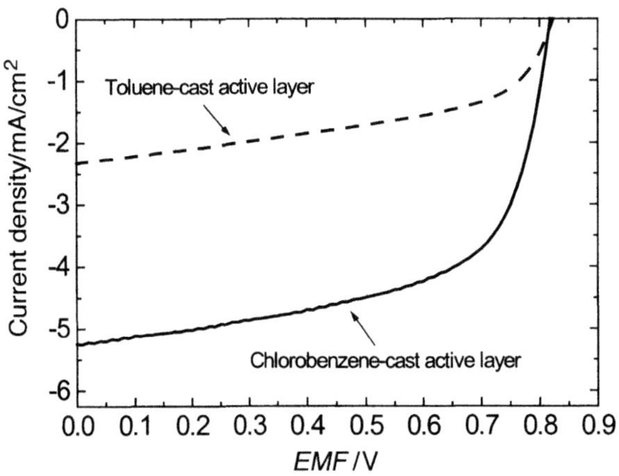

Fig. 7. Characteristics for devices with an active layer that is spin-coated from a toluene solution (dashed; $J_{SC} = 2.33\,\text{mA/cm}^2, V_{OC} = 0.82\,\text{V}, FF = 0.50, \eta_{AM1.5} = 0.9\%$) and from a chlorobenzene solution (full line; $J_{SC} = 5.25\,\text{mA/cm}^2, V_{OC} = 0.82\,\text{V}, FF = 0.61, \eta_{AM1.5} = 2.5\%$); data refer to devices illuminated with an intensity of $80\,\text{mW/cm}^2$ with an AM 1.5 spectral mismatch factor of 0.753; the temperature of the samples during measurement was 50°C

The power conversion efficiency of bulk heterojunction solar cells can be improved dramatically by manipulating the morphology of the components as well as of the blend. The resulting several-fold enhancement in the short-circuit current of photovoltaic devices originates from improved mobility of the charge carriers [54, 55]. Figure 6 shows AFM images of the surfaces of *MDMO-PPV:PCBM* blend films spin-coated using either toluene or chlorobenzene. The images show clearly a much more uniform mixing of the constituents using chlorobenzene. A plot of the current density vs. voltage for the two devices is shown in Fig. 7. The optical absorption spectra of the active layer films are nearly identical, but the external quantum efficiency, or incident photon to converted electron (IPCE) ratio, as a function of wavelength shows a large increase [56, 57]. The hole mobility in pristine *MDMO-PPV* is approximately one order of magnitude greater in films spin-coated from chlorobenzene as compared to toluene cast ones [55].

Future Aspects

One of the limiting parameters in plastic solar cells is their mismatch to the solar spectrum. The use of low bandgap polymers ($E_g < 1.8\,\mathrm{eV}$) to expand the spectral region of bulk heterojunction solar cells can enhance the spectral photon harvesting. Further optimization of device performance can be achieved by optimization of the device physics:

i) Optimize the choice of metallic electrodes to achieve good ohmic contacts on both sides for collection of the oppositely charged photocarriers
ii) Optimize the choice of the D/A pair (the energetics determine the open circuit potential)
iii) Optimize the network morphology of the phase separated composite material for enhanced transport and carrier generation

In organic photovoltaic elements, generally two different tasks are to be clearly distinguished: The photoinduced charge generation (electron transfer efficiency) and the transport of created charges to the electrodes (charge carrier mobility). These two different tasks are expected to be fulfilled by one and the same material simultaneously. A possible strategy is to separate the two tasks by using separate components in a device for the charge transport and for the charge generation.

Acknowledgements

The authors gratefully acknowledge contributions from the co-workers at the University of Linz as well as from numerous international collaboration partners. Financial supports of the European Commission (DGXII, JOULE III), Quantum Solar Energy Linz GesmbH (Austria), *Christian Doppler* Society (Austria), the *Magistrat Linz*, the *Land Oberösterreich* (ETP), and the Austrian Science Fund (FWF P-12680CHE) are gratefully acknowledged.

References

[1] Burroughes JH, Bradley DDC, Brown AR, Marks RN, Mackay K, Friend RH, Burns PL, Holmes AB (1990) Nature **347**: 539

[2] Braun D, Heeger AJ (1991) Appl Phys Lett **58**: 1982
[3] Braun D, Heeger AJ, Kroemer H (1991) J Electron Mat **20**: 945
[4] Greenham NC, Moratti SC, Bradley DDC, Friend RH, Holmes AB (1993) Nature **365**: 628
[5] Parker ID (1994) J Appl Phys **75**: 1656
[6] Yu G, Pakbaz K, Heeger AJ (1994) Appl Phys Lett **64**: 3422
[7] Antoniadis H, Hsieh BR, Abkowitz MA, Stolka M, Jenekhe SA (1993) Polymer Preprints **34**: 490
[8] Antoniadis H, Hsieh BR, Abkowitz MA, Jenehke SA, Stolka M (1994) Synth Met **62**: 265
[9] Rieß W, Karg S, Dyakonov V, Meier M, Schwoerer M (1994) Journal of Luminescence **60–61**: 906
[10] Sariciftci NS, Smilowitz L, Heeger AJ, Wudl F (1992) Science **258**: 1474
[11] Sariciftci NS, Braun D, Zhang C, Srdanov V, Heeger AJ, Stucky G, Wudl F (1993) Appl Phys Lett **62**: 585
[12] Yu G, Gao J, Hummelen JC, Wudl F, Heeger AJ (1995) Science **270**: 1789
[13] Halls JJM, Walsh CA, Greenham NC, Marseglia EA, Friend RH, Moratti SC, Holmes AB (1995) Nature **376**: 498
[14] Smilowitz L, Sariciftci NS, Wu R, Gettinger C, Heeger AJ, Wudl F (1993) Phys Rev B **47**: 13835
[15] Kraabel B, Hummelen JC, Vacar D, Moses D, Sariciftci NS, Heeger AJ, Wudl F (1996) J Chem Phys **104**: 4267
[16] Sariciftci NS, Heeger AJ (1994) Int J Mod Phys B **8**: 237
[17] Sariciftci NS (1995) Prog Quant Electr **19**: 131
[18] Wei X, Vardeny ZV, Sariciftci NS, Heeger AJ (1996) Phys Rev B **53**: 2187
[19] Morita S, Zakhidov AA, Yoshino K (1992) Solid State Commun **82**: 249
[20] Yoshino K, Yin XH, Morita S, Kawai T, Zakhidov AA (1993) Solid State Commun **85**: 85
[21] Morita S, Zakhidov AA, Yoshino K (1993) Jpn J Appl Phys **32**: L873
[22] Yoshino K, Akashi T, Yoshimoto K, Morita S, Sugimoto R, Zakhidov AA (1994) Solid State Commun **90**: 41
[23] Halls JJM, Friend RH (1997) Synth Met **85**: 1307
[24] Yu G, Heeger AJ (1995) J Appl Phys **78**: 4510
[25] Sariciftci NS, Heeger AJ (1996) In: Nalwa HS (ed) Handbook of Organic Conductive Molecules and Polymers; vol 1. Wiley, New York, p 414–450
[26] Kraabel B, Lee CH, McBranch D, Moses D, Sariciftci NS, Heeger AJ (1993) Chem Phys Lett **213**: 389
[27] Kraabel B, McBranch D, Sariciftci NS, Moses D, Heeger AJ (1994) Phys Rev B **50**: 18 543
[28] Brabec CJ, Zerza G, Cerullo G, De Silvestri S, Luzzati S, Hummelen JC, Sariciftci NS, Chem Phys Lett (submitted)
[29] Lee CH, Yu G, Sariciftci NS, Moses D, Pakbaz K, Zhang C, Heeger AJ, Wudl F (1993) Phys Rev B **48**: 15425
[30] Allemand PM, Srdanov G, Koch A, Khemani K, Wudl F, Rubin Y, Diederich FN, Alvarez MM, Anz SJ, Whetten RL (1991) J Am Chem Soc **113**: 2780
[31] Dyakonov V, Zoriniants G, Scharber MC, Brabec CJ, Janssen RAJ, Hummelen JC, Sariciftci NS (1999) Phys Rev B **59**: 8019
[32] Meier H (1974) Organic Semiconductors. Verlag Chemie, Weinheim
[33] Tang CW (1986) Appl Phys Lett **48**: 183
[34] Simon J, Andre J-J (1985) Molecular Semiconductors. Springer, Berlin
[35] Brabec CJ, Sariciftci NS (1999) In: Hadziioannou G, van Hutten PF (eds) Semiconducting Polymers. Wiley-VCH, Weinheim, chapt 15, p 515–560
[36] Kanicki J (1986) In: Skotheim TA (ed) Handbook of Conducting Polymers, vol 1. Dekker, New York, p 543–660
[37] Kaneko M (1997) In: Nalwa HS (ed) Handbook of Organic Conductive Molecules and Polymers, vol 4. Wiley, New York, p 661
[38] O'Regan B, Grätzel MA (1991) Nature **353**: 737

[39] Bach U, Lupo D, Comte P, Moser JE, Weissörtel F, Salbeck J, Spreitzer H, Grätzel M (1998) Nature **395**: 583
[40] Yamashita Y, Takashima W, Kaneto K (1993) Jpn J Appl Phys **32**: L1017
[41] Wöhrle D, Meissner D (1991) Adv Mat **3**: 129
[42] Rostalski J, Meissner D (2000) Solar Energy Materials and Solar Cells **61**: 87
[43] Rostalski J, Meissner D (2000) Solar Energy Materials & Solar Cells **63**: 37
[44] Tsuzuki T, Shirota Y, Rostalski J, Meissner D (2000) Solar Energy Materials and Solar Cells **61**: 1
[45] Roman LS, Mammo W, Pettersson LAA, Andersson MR, Inganäs O (1998) Adv Mat **10**: 774
[46] Halls JJM, Friend RH (1997) Synth Met **85**: 1307
[47] Yu G, Zhang C, Heeger AJ (1994) Appl Phys Lett **64**: 1540
[48] Antoniadis H, Hsieh BR, Abkowitz MA, Stolka M, Jenekhe SA (1993) Polymer Preprints **34**: 490
[49] Karg S, Riess W, Dyakonov V, Schwoerer M (1993) Synth Met **54**: 427
[50] Granstrom M, Petritsch K, Arias AC, Lux A, Andersson MR, Friend RH (1998) Nature **395**: 257
[51] Greenham NC, Moratti SC, Bradley DDC, Friend RH, Holmes AB (1993) Nature **365**: 628
[52] Neugebauer H, Brabec CJ, Hummelen JC, Sariciftci NS (2000) Solar Energy Materials & Solar Cells **61**: 35
[53] Neugebauer H, Brabec CJ, Hummelen JC, Janssen RAJ, Sariciftci NS (1999) Synth Metals **102**: 1002
[54] Shaheen SE, Brabec CJ, Padinger F, Fromherz T, Hummelen JC, Sariciftci NS, Appl Phys Lett (in press)
[55] Brabec CJ, Shaheen SE, Fromherz T, Padinger F, Hummelen JC, Dhanabalan A, Janssen RAJ, Sariciftci NS, Synth Met (in press)
[56] Geens W, Shaheen SE, Brabec CJ, Poortmans J, Sariciftci NS, In: H. Kuzmany, Jörg Fink, Michael Mehring and Siegmar Roth (eds) Properties of Novel Materials. Institute of Physics (in press)
[57] Shaheen SE, Brabec CJ, Sariciftci NS, Geens W, Poortmans J (in preparation)

Received December 19, 2000. Accepted December 22, 2000

Intermolecular Order of Poly-(2,5-dimethyl-*para*-phenylene vinylene) and Poly-(*para*-phenylene vinylene) – A Comparison

Roland Resel[1,*], **Raf Kiebooms**[2], **Dirk Vanderzande**[2], and **Franz Stelzer**[3]

[1] Institut für Festkörperphysik, Technische Universität Graz, A-8010 Graz, Austria
[2] Institut voor Materiaalonderzoek, Limburgs Universitair Centrum, B-3590 Limburg, Belgium
[3] Institut für Chemische Technologie Organischer Stoffe, Technische Universität Graz, A-8010 Graz, Austria

Summary. Poly-(*para*-phenylene vinylene) (**2b**, *PPV*) and poly-(2,5-dimethyl-*para*-phenylene vinylene) (**1b**) were prepared *via* the formation of a double bond by thermal elimination of an octylsulfinyl side group at 200°C under vacuum. X-Ray diffraction experiments revealed a certain degree of long-range order within both polymers. However, considerable differences in the diffraction pattern were observed, the packing of the polymer chains within the bulk structure being responsible for these differences. Within the crystal structure of **2b**, neighbouring *PPV* chains pack in a zig-zag arrangement (herringbone pattern). Analysis of the diffraction pattern of **1b** on the basis of the crystal structure of a model compound (2,5,2′,5′-tetramethylstilbene) revealed that the molecular planes of neighbouring polymer chains pack parallel to each other. A model for the crystal structure of **1b** is given. The change of packing from a herringbone to a parallel arrangement of the molecular planes is related to the introduction of methyl groups as side chains to *PPV*.

Keywords. Poly-(*para*-phenylene vinylene); *PPV*; X-Ray diffraction; Crystal structure.

Introduction

Poly-(*para*-phenylene vinylene) (*PPV*) and its derivatives are semiconducting, π-conjugated polymers with high potential for future applications [1]. New display technology based on electroluminescence of these materials is an ongoing subject of industrial research. Since pure *PPV* is insoluble in conventional solvents, two methods have been developed to render this material processible for technological use. One way is the coupling of flexible side chains to the *PPV* polymer backbone, poly-(2-methoxy-5-(2′-ethyl-hexyloxy)-*para*-phenylene vinylene), denoted as *MEH-PPV*, being the most prominent among these soluble *PPV* derivatives [2, 3]. The second way is the use of soluble precursor polymers which can be converted by thermal treatment to *PPV* or *PPV* derivative polymers [4].

The amorphous state is preferred for technological use of *PPV* derivative polymers [5]. Depending on the preparation procedure and on the type of side

* Corresponding author

chains, the polymer is either amorphous or semicrystalline [6]. Electrical conductivity as well as optical properties depend on the arrangement of neighbouring polymer chains [7, 8]; the orientation of neighbouring *PPV* chains relative to each other and the distance between them are of particular importance [9].

PPV crystallizes in a monoclinic symmetry (space group $P2_1/c$) with lattice constants $a = 8.30, b = 6.05, c(\text{chain axis}) = 6.58\,\text{Å}$, $\beta = 123°$ [10]. The atoms of one polymer chain are arranged approximately in one molecular plane. The molecular planes of neighbouring polymer chains are ordered in a zig-zag arrangement of neighbouring molecules (herringbone pattern), they enclose a tilt angle between 64 and 82° relative to each other [10, 11]. Few experimental data are available for *PPV* derivative polymers; *e.g.*, lattice constants of *MEH-PPV* have been determined [12]. However, the relative arrangement and the distance between the *PPV* backbones are not known. Single crystal investigations on oligomers of *PPV* with attached side chains reveal that the side chains influence the packing of *PPV* backbones considerably [13, 14]. The implementation of side chains promotes the tendency to separate the backbones from each other.

The aim of this study is to characterize two different type of *PPV* polymers which prepared *via* an identical precursor route. The only difference between both *PPV* polymers is the introduction of methyl groups at positions 2 and 5 of the phenyl ring, changing the features of the crystal structure of the polymer drastically.

Results and Discussion

The diffraction pattern of **1b** is given in the upper part of Figure 1. **1b** shows two broad peaks located at $2\Theta = 14.5°(d = 6.1\,\text{Å})$ and at $2\Theta = 25.0°(d = 3.6\,\text{Å})$. The positions of these two diffraction peaks agrees quite well with transmission electron diffraction data on methyl-substituted *PPV* ($d = 6.5$ and $3.6\,\text{Å}$ [20]).

The diffraction pattern of **2b** is given in the lower part of Fig. 1. **2b** shows only one peak at $2\Theta = 20.8°(d = 4.3\,\text{Å})$ and a shoulder at $2\Theta = 27.7°(d = 3.2\,\text{Å})$. Bars on the bottom line in Fig. 1 give the 2Θ-positions and the relative intensities of the main diffraction peaks in crystalline *PPV* ($2\Theta = 20.9°(d = 4.24\,\text{Å})$, $2\Theta = 22.3°(d = 3.97\,\text{Å})$, and $2\Theta = 28.6°(d = 3.12\,\text{Å})$ [11]). A comparison of the diffraction pattern of **1b** with the peak characteristics of crystalline *PPV* leads to the conclusion that **1b** is partly crystalline. The peak intensities of **2b** differ from the expected ratios of *PPV.* Two possible explanations can be given: *i*) within the film, the crystallites of **2b** display a preferred orientation or *ii*) an amorphous part of **2b** exists which would result in a broad peak at about $2\Theta = 20°(d = 4.4\,\text{Å})$ [11].

Although the diffraction patterns of **1b** and **2b** show peaks with a characteristic peak width of several degrees, we conclude that both polymers show a certain degree of long-range order. The experimental diffraction patterns of **1b** and **2b** differ considerably from each other. Obviously, the crystal structure properties of the two polymers are not comparable. Generally, a crystal structure of a molecular material is determined by the conformation of the polymer as well as by the packing of the polymer chains. Therefore, the conformation and packing of **1b** and **2b** have to be analyzed. The results on **2b** are interpreted on basis of crystalline *PPV* [11]. The diffraction pattern of **1b** is analyzed by a comparison with the model substance 2,5,2′,5′,-tetramethylstilbene [21].

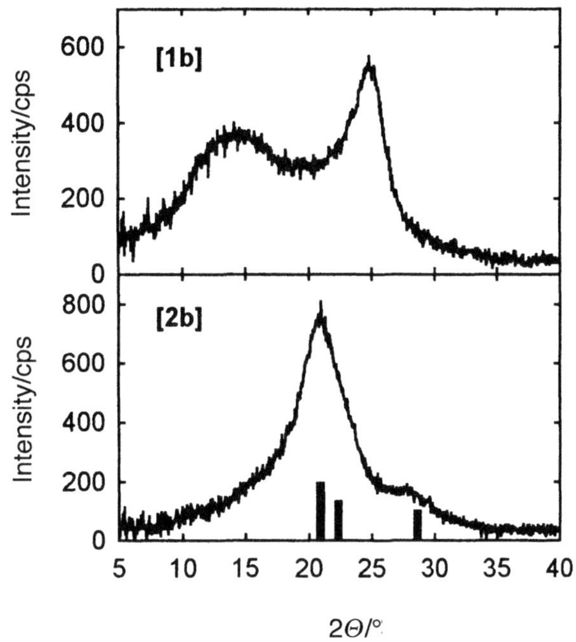

Fig. 1. $\Theta/2\Theta$ – diffraction pattern of **1b** (above) and **2b** (below); the characteristic diffraction peaks of *PPV* with respect to their position and relative intensities are given by bars in the bottom line [11]

Crystal structure studies of *PPV* and of stilbene, a model compound for *PPV*, reveal that the phenyl rings are in *trans*-arrangement to the vinyl unit (*trans*-conformation) [11, 13, 14]. Within a *cis*-arrangement, the adjacent phenyl rings are drastically twisted relative to each other [22]. The *trans*-conformation of the molecules can be characterized on the basis of *PPV*: the neighbouring phenyl rings within one polymer chain are in the same plane, whereas the dihedral angle between the phenyl ring and the vinyl segment is about 10° [11]. It can be concluded that within one polymer chain the atoms are arranged approximately within one molecular plane. The packing of the plain polymer chains to a crystal structure appears in a herringbone pattern. Figure 2 gives the herringbone pattern of *PPV*. The polymer backbones are projected along the polymer chain axes, hydrogen atoms are not depicted. Three different types of net planes which give high intensity reflections in X-ray diffraction experiments are indicated by dotted lines. Since there is a good agreement in the diffraction pattern of our *PPV* and literature data (Fig. 1), the interplanar distances are taken as 4.24, 3.97, and 3.12 Å according to Ref. [11]. Comparison of the crystal structure properties of *PPV* with that of stilbene reveal that the conformation as well as the molecular packing is similar [23]. However, *PPV* and stilbene have different tilt angles of neighbouring molecules: whereas the stilbene molecules are tilted by 60° relative to each other, the tilt angle of adjacent *PPV* molecules is slightly larger.

In contrast to *PPV*, the polymer **1b** has additional methyl groups attached to the *PPV* backbone at positions 2 and 5 of the phenyl ring. The conformation of **1b** within the bulk is deduced from the model substance 2,5,2′,5′-tetramethylstilbene [21]: it is known that the two phenyl rings of the model substance are in one plane;

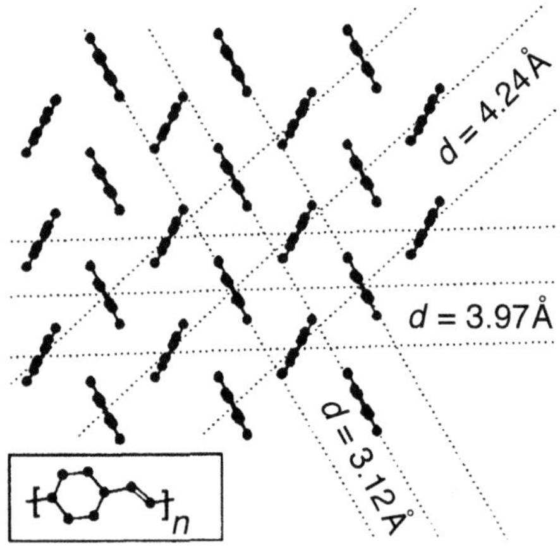

Fig. 2. Herringbone-type packing of *PPV* chains within the crystal structure; the polymer chains are projected in direction of their long axes; the three indicated planes cause the three characteristic high intensity reflections of *PPV*; the inset on the bottom/left side gives the chemical structure of *PPV*

however, the dihedral angle between the phenyl ring and the vinyl segment is 26°. The packing of the model substance within the crystal structure occurs in a parallel arrangement of the molecular planes of neighbouring molecules. The distance between two neighbouring molecular planes is 3.48 Å, the distance between molecular planes of lateral adjacent molecules is 7.7 Å. A similar molecular arrangement is suggested for the crystal structure of **1b** which is depicted in Fig. 3. The molecules are drawn schematically in a projection along the long molecular axes. The net plane between the molecular planes is 3.6 Å, which agrees quite well with the model substance (3.48 Å), a typical distance of two π-stacked aromatic

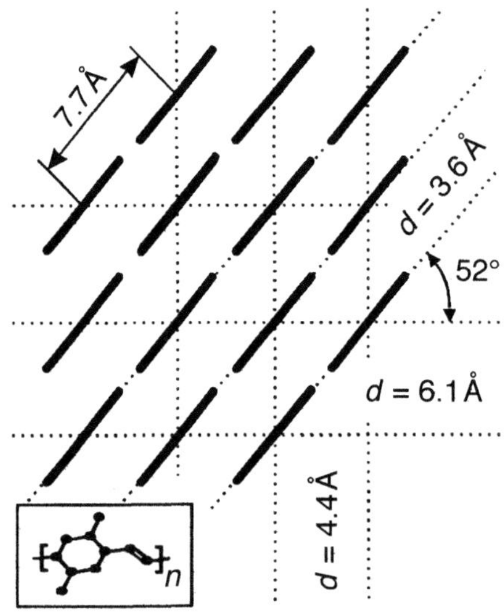

Fig. 3. Model structure of the methyl substituted *PPV* polymer **1b**; the molecules are drawn schematically in direction of their long axes; the distance between two adjacent molecules is 7.7 Å, the distance between two molecular planes 3.6 Å; the three indicated net planes cause high-intensity diffraction peaks; the inset on the bottom/left side gives the chemical structure of **1b**

planes as observed in crystal structures of aromatic molecules [24]. The resulting peak of these net planes show the highest intensity in the diffraction pattern of the model compound as well as of **1b**. One further set of net planes of the model structure which should result in high intensity X-ray diffraction peaks is at 6.1 Å ($2\Theta = 14.5°$). This peak is clearly visible in the diffraction pattern of **1b**. The diffraction peak corresponding to the interplanar distance of 4.4 Å ($2\Theta = 20.2°$) is not clearly visible in the diffraction pattern of **1b**. However, due to the general large peak width, this peak is probably superimposed.

The lateral packing of adjacent molecules is determined by the attached volume of the methyl groups. The lateral distance between adjacent polymer chains is 7.7 Å. This small distance is only possible when the lateral adjacent molecules are shifted along the polymer chain of about a half of the monomer unit length. This shift results in a close packing of adjacent molecules. The lattice constants of the model structure are $a = 6.1, b = 8.8$, and $c = 6.7$ Å. The shift of adjacent molecules along the polymer backbone causes a doubling of the interplanar distance of 4.4 Å to the lattice constant $b = 8.8$. The lattice constant c (along the polymer backbone) is the length of one monomer unit of the polymer **1b**. Within this unit cell, the molecular planes are tiled 50° relative to the *bc*-plane.

The herringbone packing as well as the parallel stacking is generally observed in polynuclear aromatic hydrocarbons of long, thin shape; molecules of larger shape tend to arrange their molecular planes relative to each other [25, 26]. Since it is known that extended molecules tend to adopt a parallel arrangement [24], it can be assumed that in several types of side chain containing *PPV* the π-conjugated molecular planes prefer parallel packing.

Conclusions

Poly-(*para*-phenylene vinylene) **2b** and poly-(2,5-dimethyl-*para*-phenylene vinylene) **1b** were prepared from two slightly different precursor materials and by the same route. The resulting polymers are nearly identical, the difference being only the introduction of two methyl groups at positions 2 and 5 of the phenyl rings. The X-ray diffraction pattern of **1b** and **2b** show different features. The diffraction pattern of **2b** can be explained on the basis of the known crystal structure. A structural model for **1b** is developed on basis of the X-ray diffraction pattern and the crystal structure of the model substance 2,5,2′,5′-tetramethylstilbene. Both polymers show different types of packing of the molecules to a crystal structure. Whereas in **2b** the polymer chains pack in the well known herringbone pattern, in case of **1b** a parallel packing of the polymer chains is observed.

Experimental

The synthesis is based on the precursor method first reported by *Louwet et al.* [15] making use of the formation of a double bond by thermal elimination of a sulfoxy side chain. The main advantages of these precursor polymers are their increased stability as compared to the standard *Wessling* route, solubility in organic solvents, and the elimination of non-corrosive materials under thermal treatment [16]. The modified procedure was used to synthesize the poly-(*n*-octylsulfinyl-2,5-dimethyl-*para*-phenylene ethylene) precursor **1a** and the poly-(*n*-octylsulfinyl-*para*-phenylene ethylene) precursor **2a** [17].

The typical procedure for the synthesis of **1a** is as follows: A solution of 0.178 mol of Na-*tert*-butoxide and 0.181 mol of 1-octanethiol in 300 cm^3 of MeOH is added to a clear solution of 67.1 g *Wessling* salt in 200 cm^3 of MeOH. The mixture is stirred for another 30 min, and solvent is removed *in vacuo*. *n*-Octane is added and removed *in vacuo* (3x). The remaining oil is dissolved in $CHCl_3$ and washed with water and $NaHCO_3$ (1 *M*). The solution is dried over $MgSO_4 \cdot H_2O_2$ and added to a solution of the crude compound together with TeO_2 in 300 cm^3 of MeOH/1,4-dioxane (5:1). After 5 h of stirring the reaction is stopped by addition of 100 cm^3 of a saturated NaCl solution. The mixture is extracted with $CHCl_3$ and dried over $MgSO_4$. Yield: 43.4 g of crude monomer. The monomer α-chloro-α'-octylsulfinyl-*p*-xylene was further purified by crystallization from $CHCl_3$/hexane. 0.106 mol of Na-*tert*-butoxide in 240 cm^3 MeOH are added at once to a mechanically stirred solution of 0.082 mol of α-chloro-α'-octylsulfinyl-*p*-xylene in 570 cm^3 *s*-butanol. After 1 h of stirring the mixture is poured in ice water, neutralized with 1 *M* HCl, and extracted with $CHCl_3$. The solvent is evaporated to remove the *s*-butanol, redissolved in $CHCl_3$, and precipitated in diethylether. The precipitated was collected and dried *in vacuo*. Yield: 14.7 g polymer.

Exactly the same procedure was followed for the synthesis of **2a**. The molecular weight of the precursors were determined by gel permeation chromatography (GPC) in THF against a polystyrene standard. **1a** has a molecular weight (M_W) of 879976 and a polydispersity (*PD*) of 3.26; **2a**: $M_W = 184115$, $PD = 2.48$. The glass transition temperatures (T_g) of the precursor polymers are 44 and 63°C, respectively.

^{1}H NMR (400 MHz, $CDCl_3$ **1a**): $\delta = 0.81, 1.18, 1.63, 1.92, 2.14, 2.25, 2.95, 3.42, 3.65, 3.88$ ppm; ^{13}C NMR (100 MHz, $CDCl_3$ **1a**): $\delta = 14.0, 18.9, 22.5, 22.8, 28.9, 31.6, 34.4, 49.5, 65.5, 129.1, 131.9, 132.4, 133.7, 134.9, 136.2, 136.7$ ppm; ^{1}H NMR (400 MHz, $CDCl_3$ **2a**): $\delta = 0.81, 1.17, 1.59, 1.96, 2.18, 2.26, 3.03, 3.62, 6.94$ ppm; ^{13}C NMR (100 MHz, $CDCl_3$ **2a**): $\delta = 14.0, 22.4, 22.5, 23.0, 28.7, 28.9, 29.0, 31.6, 36.3, 49.7, 65.2, 69.7, 128.4, 128.9, 129.4, 129.9, 131.5, 132.3, 138.2$ ppm.

The conversion to the fully conjugated system was carried out at a temperature of 200°C in vacuum. Figure 1 shows the chemical structures and the thermal conversion of the precursor materials to poly-(2,5-dimethyl-*para*-phenylene vinylene) (**1b**) and to *PPV* (**2b**). Figure 2 displays the absorption and photoluminescence spectra of the conjugated polymer **1b**. The spectra are typical for *PPV* derivative polymers [18, 19].

Films of **1a** and **2a** were prepared from a $CHCl_3$ solution by drop casting onto a silicon single crystal which was cut along a non-reflective direction. This substrate was chosen because of its low background scattering in X-ray diffraction experiments. The precursor films were converted to films of **1b** and **2b** at a temperature of 200°C in a vacuum of 10^{-3} mbar for 3 h. The thickness of the films was approximately 20 μm. After the diffraction experiments the films were lifted off the substrates. Highly flexible, good quality freestanding films were obtained.

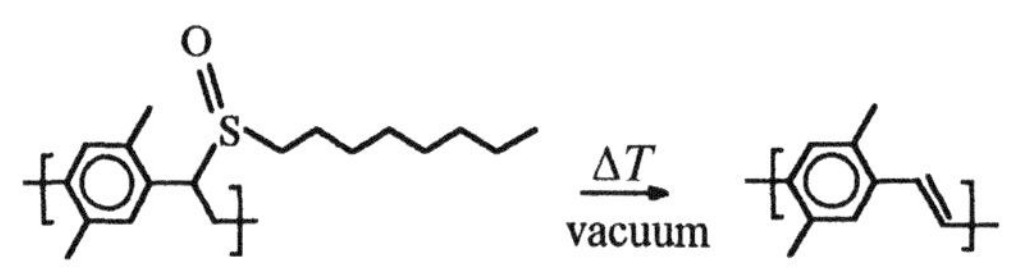

1a **1b**

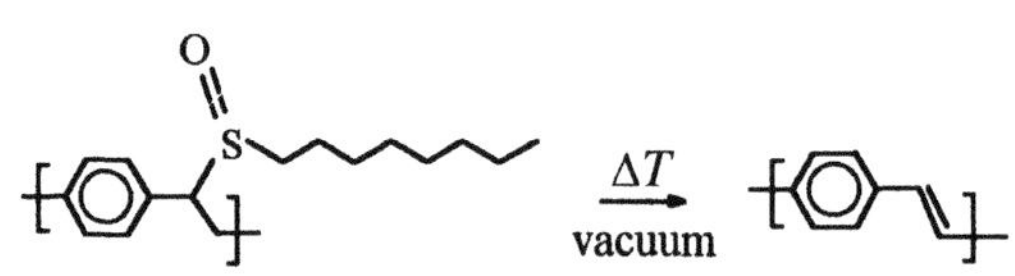

2a **2b**

Fig. 4. Chemical structures of **1a,b** and **2a,b**

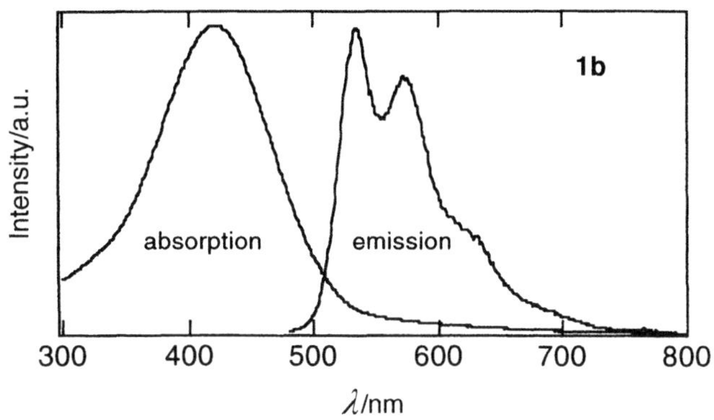

Fig. 5. Photoluminescence and optical absorption spectra of **1b** reveal the typical characteristics of *PPV* derivative polymers

X-Ray diffraction experiments were performed with a Siemens D501 diffractometer in $\Theta/2\Theta$ *Bragg-Brentano* geometry using Ni-filtered CuK_α radiation. The experimental results were corrected for the scattering of the substrate. 2Θ is the scattering angle of the diffracted beam, d gives the interplanar distance corresponding to the diffraction peak calculated on basis of *Bragg*'s equation. The software package POWDER CELL was used to calculate diffraction patterns and to visualize the crystal structures [27].

Acknowledgements

R. Kiebooms is indebted to the National Science Foundation – Flanders for financial support. *R. Resel* and *F. Stelzer* are grateful to the Austrian Science Foundation – *Spezialforschungsbereich Elektroaktive Stoffe* for the general support of their collaboration.

References

[1] Skotheim TA, Elsenbaumer RL, Reynolds JR (1998) Handbook of Conducting Polymers, 2nd ed. Dekker, New York, chapter V
[2] Burroughes JH, Bradley DDC, Brown AR, Marks RN, Mackay K, Friend RH, Burns PL, Holmes AB (1990) Nature **347**: 539
[3] Braun D, Heeger AJ (1991) Appl Phys Lett **58**: 1982
[4] Bradley DDC (1987) J Phys D **20**: 1389
[5] Son S, Dodabalapur A, Lovinger AJ, Galvin ME (1995) Science **269**: 376
[6] Resel R, Tertinek B, Tasch S, Davey A, Blau W, Hörhold HH, Rost H, Leising G (1999) Synth Met **101**: 96
[7] Gomes de Costa P, Dandrea RG, Conwell EM (1993) Phys Rev B **47**: 1800
[8] Cornil J, Heeger AJ, Bredas JL (1997) Chem Phys Lett **272**: 463
[9] Conwell EM, Perlstein R (1996) Phys Rev B **54**: R2308
[10] Granier T, Thomas EL, Gagnon DR, Karasz FE, Lenz RW (1986) J Polym Sci B Polym Phys **24**: 2793
[11] Chen D, Winokur MJ, Masse MA, Karasz FE (1992) Polymer **33**: 3116
[12] Yang CY, Hide F, Diaz-Garcia MA, Heeger AJ, Cao Y (1998) Polymer **39**: 2299
[13] Gill RE, Meetsma A, Hadziiaonnou G (1996) Adv Mater **8**: 212
[14] Hohloch M, Maichle-Mossmer C, Hanack M (1998) Chem Mater **10**: 1327
[15] Louwet F, Vanderzande D, Gelan J, Mullens J (1995) Macromolecules **28**: 1330

[16] de Kok M, van Breemen A, Carleer R, Adriaensens P, Gelan J, Vanderzande D (1999) Acta Polym **50**: 28
[17] van Breemen A, Vanderzande D, Adriaensens P, Gelan J (1999) J Org Chem **64**: 3106
[18] Halliday DA, Burn PL, Bradley DDC, Friend RH, Gelsen OM, Holmes AB, Kraft A, Martens JHF, Pichler K (1993) Adv Mater **5**: 40
[19] Hide F, Schwartz BJ, Diaz-Garcia MA, Heeger AJ (1998) SPIE **3145**: 36
[20] Martens JHF, Marseglia EA, Bradley DDC, Friend RH, Burn PL, Holmes AB (1993) Synth Met **55–57**: 449
[21] Ogawa K, Sano T, Yoshimura S, Takeuchi Y, Toriumi K (1992) J Am Chem Soc **114**: 1041
[22] Hockless DCR, Whittall IR, Humphrey MG (1996) Acta Cryst C **52**: 3222
[23] Bouwstra JA, Schouten A, Kroon J (1984) Acta Cryst C **40**: 428
[24] Kitaigorodsky AI (1973) Molecular Crystals and Molecules. New York: Academic Press
[25] Stevens B (1962) Spectrochim Acta **18**: 439
[26] Desiraju GR, Gavezotti A (1989) Acta Cryst B **45**: 473
[27] Kraus W, Nolze G (1996) J Appl Cryst **29**: 301

Received December 14, 2000. Accepted (revised) December 20, 2000

Chemical and Optical Properties of New Highly Luminescent Alternating Oligo-*m,p*-phenylenevinylenes[a]

Horst Schlick[1], **Franz Stelzer**[1,*], and **Günther Leising**[2]

[1] Institute for Chemistry and Technology of Organic Materials, Graz University of Technology, A-8010 Graz, Austria

[2] Institute for Solid State Physics, Graz University of Technology, A-8010 Graz, Austria

Summary. In this paper we report on the synthesis and characterization of a variety of polyphenylenevinylene (*PPV*) derivatives containing *para*- and *meta*-linkages in the backbones. By the *m*-linkages the conjugation lengths of the oligo-*m,p*-phenylenevinylenes were limited, thus blueshifting the UV/Vis absorption and also the photoluminescence (PL) spectra. Alkoxy substituted oligo-*m,p*-phenylenevinylenes exhibited absorption and emission spectra at higher wavelengths than oligomers without oxygen directly attached to the backbone. Replacing *p*-phenylene by 4,4′-biphenylene moieties also led to a hypsochromic shift in the optical spectra.

We chose the *Heck* coupling reaction to synthesize substituted oligomers consisting of alternating *p*-phenylenevinylene and *m*-phenylenevinylene units. They were characterized by ^{1}H NMR, ^{13}C NMR, FT IR, UV/VIS, and GPC analysis.

Keywords. Oligo-*m,p*-phenylenevinylene; *Heck* coupling reaction; Segmented conjugation; Absorption spectra; Spectroscopy.

Introduction

Modern life would be unimaginable without products of the semiconductor industry (transistors, computer chips, light emitting diodes) and the excellent material properties (low weight, high mechanical strength) of plastics. For many applications, devices based on organic polymers could be very advantageous compared to their inorganic counterparts – not only due to the prospect of cheap fabrication alone, but also because of their typical properties such as flexibility, low weight, variation of shape, *etc.* They can be produced as thin, flexible, large area devices [1] with emission colors ranging from the ultraviolet [2] to the near infrared [3] region of the optical spectrum. Such devices already seem to be very attractive for several applications reaching from low information content alphanumeric displays and backlights for liquid crystal displays (*e.g.* of notebooks), mobile phones, and wrist watches to more sophisticated ones like colored flat panel screens. Therefore, a

---

[a] Dedicated to Prof. Dr. *Klaus Hummel* on occasion of his 70th birthday

* Corresponding author

great interest in the investigation of conjugated polymers and oligomers arose due to their optical properties like photo- and electroluminescence (PL and EL) [4]. A lot of different conjugated polymers like poly-*p*-phenylene, poly-*p*-phenylenevinylene, poly-thiophene, *etc.* and derivatives thereof have been synthesized and investigated intensively.

In our group, research on conjugated polymers has been carried out for more than fifteen years. Special emphasis has been laid on the development of new materials needed as active materials [5, 6] and as transporting layers [7] in organic devices such as LEDs or similar multilayer devices. On the other hand, luminescent materials with suitable transport properties can also be used in photoconductive devices or in the reversed mode as active materials in photovoltaic devices (solar cells) [8, 9]. There is, however, a number of factors that limit the efficiency of organic polymer LEDs. In most cases, the detected emission spectrum is similar to that produced by photoexcitation, indicating that PL and EL are based on similar mechanisms. Bearing in mind the statistical probability of the generation of three times as many triplet as singlet excitons, one can practically only attain an internal EL efficiency of up to 25% of the PL efficiency [10]. As a consequence, materials with high photoluminescence quantum efficiencies in the solid state are preferred.

The formation of the maximum number of singlet excitons, however, requires a balance of charge carrier injection. The energy barrier for charge injection depends on the relative positions of the work functions of the electrodes and the HOMO-LUMO energy levels of the polymer, causing usually an imbalance of charges. Besides, the mobility of holes in most organic materials is higher than the mobility of electrons [11–13]. Exciton formation tends to take place in a relatively narrow zone near the polymer/cathode interface, where defects (quenching centres) are common (metals, in particular, quench excitons very efficiently [14]). As a consequence, the adaption of transport interlayers to the electronic requirements for optimized LEDs seems to be a promising approach. As an example, the combination of *PPV* with *PBD* (an oxadiazole derivative) as electron transporting layer resulted in an eight-fold increase of EL efficiency compared to a single layer *PPV* device [15]. The use of fluorinated interlayers reduced the switching voltage of OLEDs [7].

A problem arises from exciton diffusion to quenching sites (defects, impurities) in conjugated polymers [16]. Therefore, preparation of devices under careful exclusion of oxygen is necessary to optimize quantum yields [17]. Several approaches to improve device performances have been tried. Due to interruption of the π-conjugation, the wave function of the excited state is kept within a defined segment, which means that the motion of the excitons to quenching sites is inhibited and the PL and EL efficiencies are increased. In practice this can be achieved by introducing saturated moieties into the conjugated polymer chain [18, 19]. For *PPV* it has been shown that incorporation of conjugation breaks raises the PL efficiencies to 80%, and devices gave a brightness of 400 Cd/m^2 at 4 V with lifetimes in excess of 1400 hours [20]. Other approaches include segmentation of the π-conjugation by *m*-phenylene linkages [21–23] or synthesis of graft-copolymers bearing the chromophors in the side chains [24–26].

Quenching of the luminescence combined with a color shift can also be due to crystallization or aggregate formation [27]. Polymer blends have been successfully used in such cases to dilute the luminescent material within a matrix. *Vestweber et al.* have reported that blending of a phenyl-substituted *PPV* (*PPPV*) with polystyrene (*PS*) increased the quantum efficiency of the device by two orders of magnitude [28]. Dilution due to statistical grafting of *PPV* sidechains on phenyl rings of *PS* led to PL efficiencies of about 95% [26].

Our goal is to construct segmented copolymers containing segments for every function: active luminescence, electron or hole conduction, crosslinkers for fixation of a desired morphology, and liquid crystalline (LC) moieties for polarization of the emitted light. On the way to these multifunctional polymers it is necessary to investigate each homosystem before combining it to a multi function single chain polymer. We have already reported on the well-controlled synthesis of LC polymers [29–31], on various *PPV* derivatives with LC side chains [32, 33], and even on the ADMET (acyclic diene metathesis) polymerization towards *PPV*s [6, 34].

With all these experiences in mind, now the combination of *para-* and *meta-*phenylenes should be investigated. We want to describe a material consisting of 1,3- and 1,4-phenylene and vinylene units that is expected to combine some of the desired properties. We focused on colortuning by segmentation of the π-conjugation of poly-*p*-phenylenevinylene derivatives by introducing *m*-linkages. This reduction of conjugation length led to a hypsochromic shift of the UV/Vis absorption and emission spectra. Thus, blue light emitting *PPV* derivatives could be achieved.

Additional advantages of polymers with controlled segmented conjugation are their high PL and EL quantum efficiencies [21–34]. Depending on the chosen polymerization reaction it is possible to achieve either statistical or alternating copolymers. The latter ones have the advantage that they contain segments of defined conjugation length, whereas statistical copolymers contain statistical distributions of conjugated segments. Most recently, published examples of alternating poly-*m,p*-phenylenevinylenes have been synthesized by the *Wittig* reaction [21–23]. In this paper we report on oligomers achieved by the *Heck* reaction, a Pd^0 catalyzed coupling reaction [35–37]. 1,4-Divinylbenzene (**1**), 4,4′-divinylbiphenyl (**2**), 2,4-dibromo-1-decyloxybenzene (**3a**), 2,4-dibromo-1-pentyloxybenzene (**3b**), and 1,3-dibromo-5-decyloxymethylbenzene (**4**) were used as monomers.

Results and Discussion

The vinylic compounds (**1, 2**) were copolymerized with the bromo compounds (**3a, b, 4**) to obtain the oligo-*m,p*-phenylenevinylenes **P1**-**P4** (Schemes 1 and 2). First PL measurements showed that the substances purified by repeated precipitation contained undefined rests of the catalytic system and side products that quenched

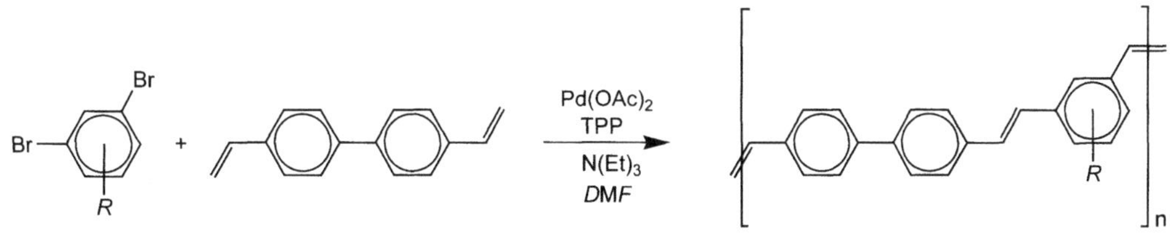

Scheme 1. Synthesis of oligomers; $R = o\text{-}(OC_{10}H_{21})$, $o\text{-}(OC_5H_{11})$, $m\text{-}(CH_2OC_{10}H_{21})$

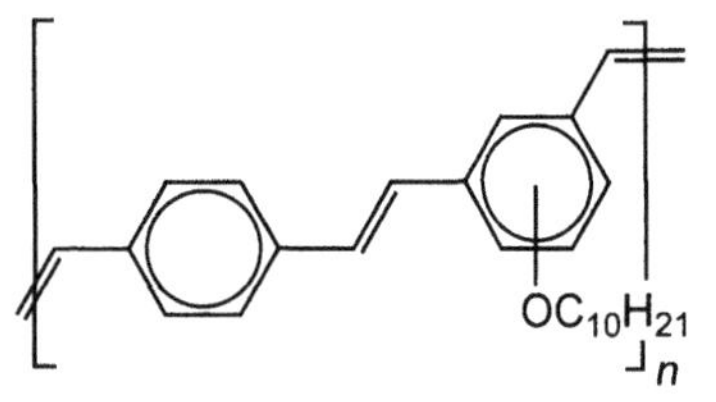

P1

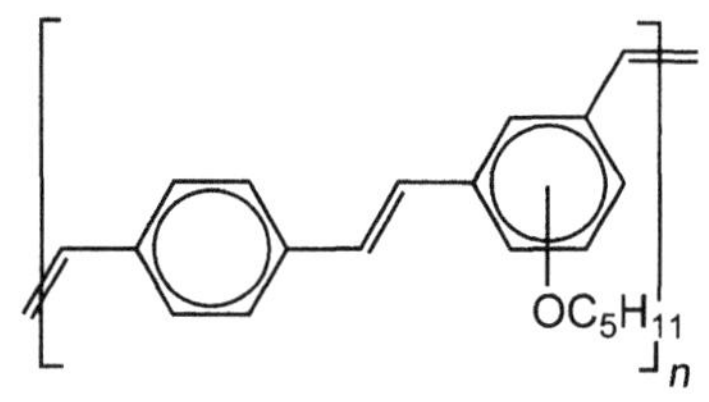

P2

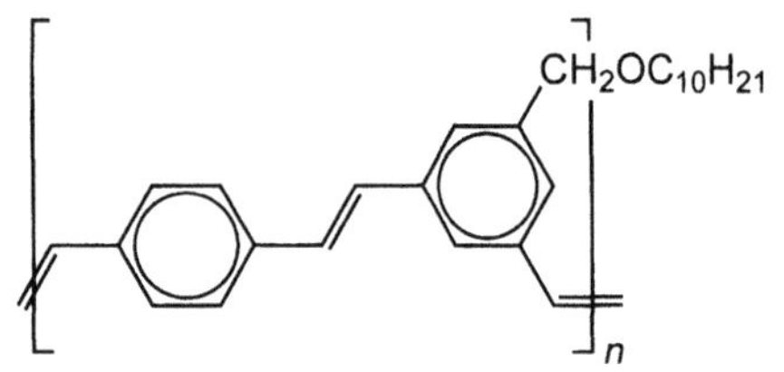

P3

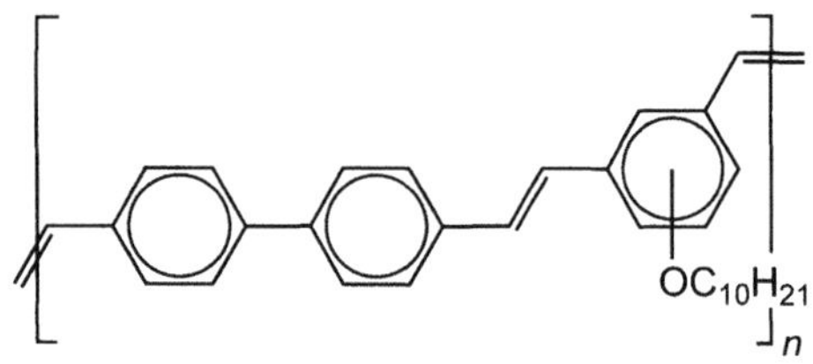

P4

Scheme 2. Overview of synthesized oligomers

PL. Further purification by column chromatography using CH_2Cl_2 as eluent was necessary to remove these substances. The obtained products were characterized by ^{1}H NMR, ^{13}C NMR, FT IR, and GPC. UV/Vis and PL measurements were performed for optical characterization.

As can be seen form the GPC plots the weight averages ranged from about 1200 (**P4**) up to 3000–4000 (**P1**-**P3**) (Table 1, Fig. 1). This means that **P4** containing the biphenylene moiety consisted of only 2–3 repeating units. The weight averages of **P1**-**P3** corresponded to about 10 repeating units including 20 phenylene rings in the backbone. Nevertheless, **P1**-**P3** contained also mass fractions consisting of only one and two repeating units. There are two possible explanations for the low molecular weights: *i*) the *Heck* reaction progressed quite slow (up to one week reaction time; compare Ref. [38] and references cited therein) and *ii*) the solubility of conjugated polymers is rather limited, which means that polymers precipitate when the molecular weights reach higher values. In our case, **P1**, **P2**, and **P3** contained one alkyloxy or alkyloxymethyl side chain per every second phenylene unit; **P4** contained only one side chain per every third phenylene unit. We observed that **P2** and **P4** precipitated when $CDCl_3$ was used for preparing samples for NMR measurements (concentrations of about 30–40 mg/cm^3) due to π-stacking. Taking into consideration that the molecular weights of **P1** and **P3** were in the same region

Table 1. Results of GPC analyses

Polymer	M_w/g · mol^{-1}	M_n/g · mol^{-1}	PDI[a]
P1	3230	1990	1.6
P2	3670	2050	1.8
P3	3710	2160	1.7
P4	1230	850	1.4

[a] PDI: polydispersity index

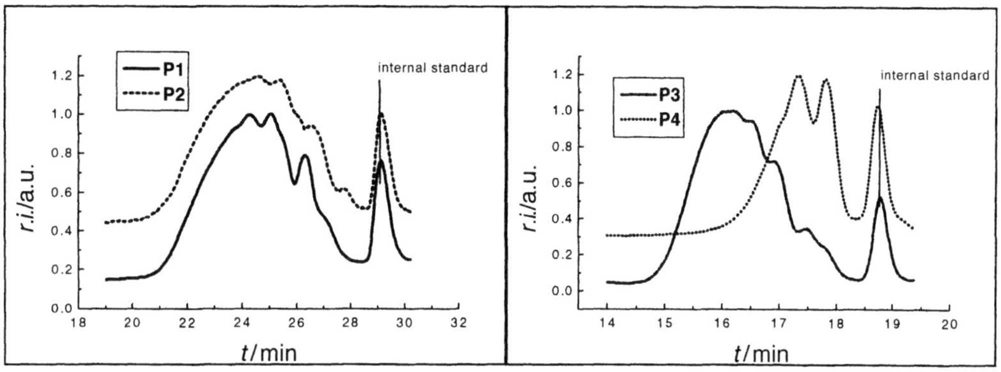

Fig. 1. GPC chromatograms of **P1-P4**

as that of **P2**, one can conclude that the intrinsic maximum molecular weight to obtain soluble **P1** and **P3** has not been reached by this reaction in the applied time. This statement is confirmed by results in connection with a polymer synthesized by the *Wittig* reaction. This polymer differed from **P1** only by the fact that it contained a higher amount of *cis* double bonds [39]. In this case we observed the above mentioned precipitation from solution in connection with a molecular weight (M_W) of $6700\,g \cdot mol^{-1}$ and a PDI of 1.9. The solubility of the oligomers increased with the number of substituents per phenylene units. This fact explains the low molecular weight of **P4** compared to that of the other ones. The properties of **P4** suffered from easy crystallization, and it was difficult to prepare transparent films. **P1** with decyloxy sidechains exhibited improved solubility and film formation properties when compared to the analogous **P2**, which contained the shorter pentyloxy substituents.

We did not focus on maximizing the molecular weights of **P1** and **P3** by, for instance, changing the reaction conditions or the leaving groups because of three reasons:

a) We did not expect a dependence of the optical (UV/Vis absorption and PL/EL emission) properties on increasing the molecular weights. The substances consisted of segments with defined conjugation length – although there was the possibility of head-to-head, head-to-tail, and tail-to-tail orientation in connection with **P1** (**P2** and **P4**). Previously published polymers with segmented conjugation due to *m*-linkages exhibited only small bathocromic shifts of about 10–15 nm in comparison with oligomers representing the same chromophor [22] (for the reason of this shift, see below).

b) We wanted to keep the substances in the range of good solubility, making processing is still possible.

c) We intend to synthesize copolymers from the oligomers still containing vinylic end groups, where the other block shall consist of various functionalized norbornene derivatives, *via* metathesis reactions.

Characterization

^{1}H NMR and IR measurements proved the oligomers to consist of nearly exclusively *trans*-vinylene moieties. *Geise et al.* have reported the *cis*-vinylene protons of the *cis,cis*-isomer of 2,5-dimethoxy-1,4-*bis*-(2-(3,4,5-trimethoxyphenyl)-ethenyl)-benzene, which was taken as a model compound for alkoxy substituted *PPV*, to be found at 6.64 and 6.55 ppm, and those of the *cis,trans*-isomer at 6.69 and 6.58 ppm [40]. We did not observe resonances in the region between 6.5 and 6.8 ppm except some small peaks which could easily be attributed to protons of terminal vinylic groups. The resonances corresponding to *trans*-vinylene protons are reported to be found at 7.06 and at 7.36 ppm for the *trans,trans*-isomer of the above mentioned model compound. We observed resonances at 7.05–7.15 and, additionally, at 7.35–7.45 for **P1**, **P2**, and **P4**. It has to be mentioned at this point that it was difficult to distinguish between the aromatic and the *trans*-vinylene protons due the irregular structures of these oligomers, which led to a variety of different signals in this region. *Cis*- and *trans*-stilbene were taken as model compounds to determine the vinylene configuration of **P3**. The expected signals of the *trans*-vinylene units (compared to those of *trans*-stilbene which are shifted to about 7.1 ppm) were observed at 7.0–7.2 ppm.

IR measurements confirmed these results. All spectra showed strong IR absorptions corresponding to *trans*-vinylene CH out-of-plane vibrations at about 960–965 cm^{-1}. In the region between 670 and 730 cm^{-1}, where the *cis*-vinylene CH out-of-plane band was expected, bands attributed to ring deformations of the aromatic rings at 690–700 cm^{-1} and weak bands at 720–730 cm^{-1} attributed to CH_2 rocking vibrations were observed.

Heitz et al. [37] have reported the possible formation of 1,1-diphenylethene by reaction of styrene with bromobenzene. They proved the presence of this compound and of 1,1-diolefinic moieties in *PPV* prepared *via* the *Heck* route by ^1H NMR. We observed broad multiplets at 5.5 ppm, which were attributed to 1,1-diolefinic moieties also by comparison with the spectrum of 1,1-diphenylethene [41]. Taking the intensity of these signals into consideration, the maximum amount of these structural defects was about 2–5%. This means that the influence on the absorption spectra could be neglected. The emission spectra were not influenced, because the emission generally occurs from those sites with the lowest HOMO-LUMO gap, *i.e.* with the longest conjugated sequences.

^{13}C NMR spectra were recorded from **P1** and **P3**, both exhibiting a variety of signals in the aromatic region due to the molecular weight distribution ranging from very low oligomers to short polymers. The spectrum of **P3** consisted of distinct sets of resonances, underlining the fact that its structural regularity was higher due to the symmetrical substitution of **4**.

Optical properties

We observed UV/Vis absorption maxima in the range of 350 to 367 nm (Fig. 2). They were blueshifted compared to those of previously published poly-*m,p*-phenylenevinylenes [21–23]. This result was due to the reduced number of substituents per repeating unit in our oligomer systems. The previously published poly-*m,p*-phenylenevinylenes contained two alkoxy substituents per every second phenylene unit. Our oligomers contained only one substituent per every second (**P1**-**P3**) or third (**P4**) phenylene unit. It is well known [12, 13] that electron donating substituents (like alkoxy groups) influence the electronic structure of the

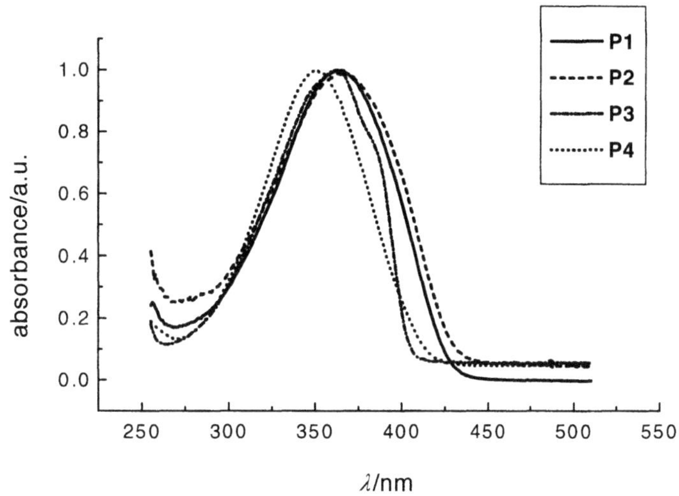

Fig. 2. UV/Vis spectra of **P1-P4** (CH_2Cl_2)

polymer backbone. They cause a decrease in the gap between the HOMO and the LUMO energy levels of the polymer chain. **P4** exhibited the lowest absorption maximum. Reasons for that result are *i*) the low amount of substituents per repeating unit and *ii*) the non-planarity of the two phenyl rings of the biphenylene unit which also reduces π-conjugation (there is a twist angle of about 23° between the phenyl rings in poly-*p*-phenylene [42]). The very low molecular weight could also have an influence. Although conjugation is reduced at *m*-phenylene moieties very effectively, there was a small bathocromic shift in the UV/Vis absorption spectra of poly-*m*,*p*-phenylenevinylenes compared to those of analogous oligomeric model compounds [22]. We explain this with the fact that the *m*-linked chromophors in polymers or higher oligomers can be compared to phenylenevinylene substituted chromophors. Therefore, the real chromophors in the polymers differ from the distyrenebenzene derivatives used as model compounds. In the case of the very short **P4** consisting of an average of only 2–3 repeating units, the relative amount of end groups leading to incomplete chromophors highly influences the absorption properties.

The differences between the spectra of **P1** and **P2** were not significant, which was not very surprising due to their similar structure. They were broader than those of **P3** and **P4**. **P3** showed a UV/Vis absorption band with a very steep slope at the long-wavelength end. It consisted of a maximum at 364 nm and vibronic shoulders at 350 and 380 nm. The maximum was in the same region as those of **P1** and **P2**, but the onset of the band was blueshifted by 28 nm, indicating a higher HOMO-LUMO energy gap.

PL measurements of the polymers were performed in solution and of thin films on quartz plates prepared by drop casting (Table 2). The visual impression of the PL emission of the polymers dissolved in CH_2Cl_2 was blue (**P1**, **P2**) and violet (**P3**, **P4**). The corresponding λ_{max} values lay at 441 (**P1**, **P2**), 428 (**P3**), and 405 nm (**P4**), respectively (Fig. 3). In addition, the spectra showed various vibronic shoulders. The solid state emission gave a bright blue (**P3**, **P4**) and blue-green (**P1**, **P2**) impression. The spectra were redshifted in comparison to those obtained from

Table 2. Results of UV/Vis absorption and PL measurements

		P1	P2	P3	P4
Solutions	$\lambda_{max,abs}$/nm[a]	367	367	364	350
(CH_2Cl_2)	$\lambda_{max,em}$/nm[b]	441	441	428	405
	λ_{exc}/nm[c]	360	360	360	340
Films	$\lambda_{max,em}$/nm[b]	476	476	462	466
(quartz plates)	λ_{exc}/nm[c]	401	407	400	400
	$\lambda_{max,exc}$/nm[d]	441	438	422	441
	observed color	blue-green	blue-green	blue	blue

[a] Wavelengths of the maxima of the UV/Vis absorption spectra; [b] wavelengths of the maxima of the PL spectra; [c] excitation wavelengths for PL measurements; [d] maxima of the excitation spectra recorded at 480, 475, 500, and 470 nm for **P1**-**P4**

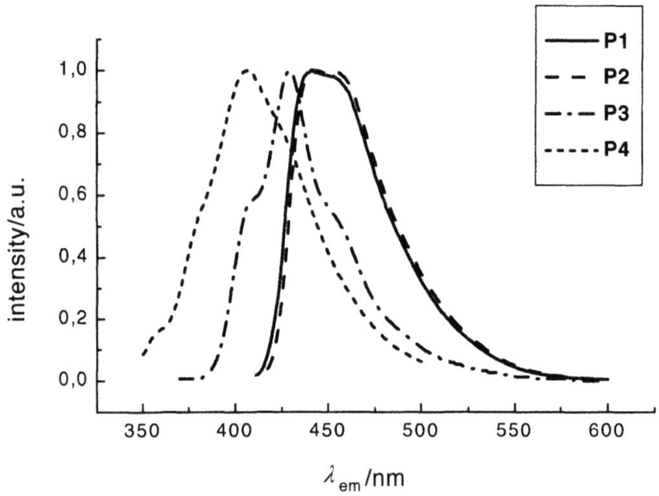

Fig. 3. PL spectra of solutions of **P1**-**P4** in CH_2Cl_2; for details, see Table 2

solutions due to effects of solid state packing. The highest shift was observed in connection with **P4**. We were not able to prepare good transparent films from this polymer – it tended to crystallization. The best film qualities were obtained from **P1** and **P3**. The $\lambda_{max,em}$ values of the solid state spectra were found at 476, 476, 463, and 466 nm for **P1**-**P4** (Fig. 4). Both the solid state spectra and those recorded from solutions showed the same trends as the absorption spectra. Replacing the *p*-phenylene moiety by a 4,4′-biphenylene unit led to a hypsochromic shift. In the case of the emission spectra, the influence of the low molecular weight should be of less importance, because emission generally occurs from sites with the lowest HOMO-LUMO gap. **P3** also emitted at lower wavelengths than **P1** and **P2**. With these compounds, the reason was found in the type of the substituent, the oxygen atom not being directly bonded to the conjugated system and thus reducing the electron donating effect.

We did not observe a dependence of the wavelength of the emission on the excitation wavelength for **P1**-**P4**, with the exception that more higher vibronic

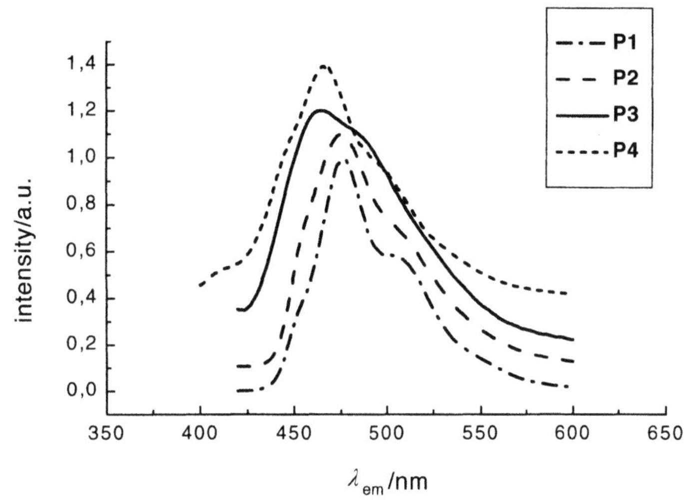

Fig. 4. PL spectra of thin films of **P1**-**P4** dropcast on quartz plates; for details see Table 2

features were observed upon setting the excitation wavelength into the onset of the emission band. Generally, the excitation wavelengths were chosen with respect to comparable optical densities.

Conclusions and outlook

We synthesized a variety of different *PPV* derivatives containing monosubstituted *m*-phenylene units. We observed a dependence of the solubility on the degree of substitution and the size of the substituents. In the case of the better soluble oligomers (**P1** and **P3**) we did not focus on maximizing the molecular weights, because we did not expect a significant influence of the higher molecular weight on the optical and optoelectronic properties and we wanted to keep them well soluble for further reactions. These materials contain vinylic end groups which will be converted, for instance *via Grubbs* type Ru-initiators, into macroinitiators, which will be used for synthesizing block-copolymers.

The UV/Vis absorption and PL spectra showed a dependence on the type of the substituent attached to the backbone. Replacing *p*-phenylene by 4,4′-biphenylene moieties turned out to be a versatile tool to influence the color of the emission. In this way we were able to synthesize new *PPV* derivatives exhibiting solid state emission in the blue and blue-green region of the spectrum. Because of the promising results in connection with photoluminescence and film formation, **P1** and **P3** are considered good candidates for the preparation of OLEDs.

Experimental

General

The FT IR spectra were obtained on a BOMEM M 100 FT IR spectrometer, ^{1}H and ^{13}C NMR spectra on a Varian Gemini 200 machine (^{1}H: 200 MHz, ^{13}C: 50 MHz). UV/Vis spectra were recorded with a HP 8452A diode array spectrophotometer, photoluminescence spectra with a Shimadzu RF-5301PC spectrofluorimeter. Mass spectra were run on a Kratos Profile mass spectrometer. Size exclusion chromatograms were detected with a Waters 410 differential refractometer (combinations of three PSS SDV gels 5 μm 8 300 mm columns with 10^4, 500, and 100 Å pore diameter (**P1**, **P2**) or two PL gel 5 μm mixed-C 7.5 300 mm columns (**P3**, **P4**) were used; eluent: $CHCl_3$, 1 cm^3/min, room temperature, calibrated with narrow distributed polystyrene standards from PSS Mainz).

Toluene, *THF*, and ether were dried over Na/K and distilled under argon, CH_2Cl_2 was dried over CaH_2 and distilled under argon, *DMF* was dried over a molecular sieve (3 Å) under argon and distilled under reduced pressure.

The chemicals were purchased from Fluka Chemie AG (terephthalaldehyde, methyltriphenyl phosphoniumbromide, 2,6-di-*t*-butyl-4-methylphenol, *BHT*, 4,4′-dibromobiphenyl, sodium hydride, 2,4-dibromophenol, *LAH*, iodomethane, copper cyanide), Aldrich Chemical Co. (3,5-dibromobencoic acid, diisobutylaluminum hydride), and Merck KGaA (potassium carbonate).

Syntheses

The monomers were synthesized according to procedures previously described in literature for the preparation of similar substances [43–46] and characterized by ^{1}H NMR, ^{13}C NMR, FT IR, and MS. 2–5% *BHT* were added to the divinylic monomers as a stabilizer.

In a typical polymerization reaction, 5.9 mmol of triethylamine were added to an equimolar (1.27 mmol) mixture of the divinyl and the dibromo monomers in an N_2-glove box. Afterwards, a suspension of 23 mg (0.102 mmol) palladium acetate and 60 mg (0.229 mmol) triphenylphosphine in 5 cm^3 *DMF* was poured into the monomer mixture. After 2(**P3**), 3(**P4**), or 6(**P1**, **P2**) days of reaction at 100°C, the dark brown suspension was cooled to room temperature, and insoluble compounds consisting of precipitated Pd, precipitated oligomers (**P2**, **P4**), and a part of the triethylammoniumbromide were removed by filtration. 50 cm^3 of $CHCl_3$ were added to the filtrate, and the solution was extracted with 50 cm^3 2 *M* HCl to remove the rest of the ammonium salts and the *DMF*. The organic phase was washed with saturated $NaHCO_3$ solution, dried over Na_2SO_4, and reduced to 2–3 cm^3. The solution was dropped into methanol three times to precipitate the oligomers. Purification by column chromatography (silicagel, CH_2Cl_2 as eluent) was necessary to obtain yellow (**P1**, **P2**), light-green (**P3**), or brownish (**P4**) compounds.

The yields of the soluble fractions were between 28 (**P4**) and 58% (**P3**).

Oligo-(1,4-phenylenevinylene-co-4-decyloxy-1,3-phenylenevinylene) (**P1**)

^{1}H NMR ($CDCl_3$): $\delta = 7.80$–6.64 (arom., vinylene, terminal vinyl (α-H)), 5.78 (d, terminal vinyl (*cis*-β-H), 5.27 (d, terminal vinyl (*trans*-β-H)), 4.06–3.95 (-OCH_2-), 2.02–1.10 (methylene), 0.99–0.87 (methyl) ppm; ^{13}C NMR ($CDCl_3$): $\delta = 156.5$, 137.7, 137.5, 130.2, 129.5, 128.7, 127.3, 124.2, 123.3, 120.5, 113.8, 112.4 (arom., vinylene, terminal vinyl), 69.5, (-OCH_2-), 32.2–22.4 (methylene), 14.2 (methyl) ppm; FT IR (NaCl): $\nu = 3053$, 3018 (CH stretch, vinylene, vinyl, arom.) 2920, 2852 (CH stretch, aliph.), 1625, 1592, 1513 (C=C stretch, vinylene + arom.), 1465, 1256, 1100 (C–O stretch), 1018, 964 (CH out-of-plane, *trans*-vinylene), 812 (CH out-of-plane, arom.), 754, 692 (ring deformation, arom.), 540 cm^{-1}.

Oligo-(1,4-phenylenevinylene-co-4-pentyloxy-1,3-phenylenevinylene) (**P2**)

^{1}H NMR ($CDCl_3$): $\delta = 7.85$–6.65 (arom., vinylene, terminal vinyl (α-H)), 5.74 (d, terminal vinyl (*cis*-β-H)), 5.27 (terminal vinyl (*trans*-β-H)), 4.20–3.75 (-OCH_2-), 1.80–1.10 (methylene), 1.05–0.55 (methyl) ppm; FT IR (NaCl): $\nu = 3017$ (CH stretch, vinylene, vinyl, arom.) 2953, 2926, 2866 (CH stretch, aliph.), 1627, 1596, 1515, (C=C stretch, vinylene + arom.), 1493, 1467, 1251, 1110 (C–O stretch), 964 (CH out of plane, *trans*-vinylene), 811 (CH out-of-plane, arom.), 733 (CH_2 rocking), 692 (ring deformation, arom.) cm^{-1}.

Oligo-(1,4-phenylenevinylene-co-5-decyloxymethyl-1,3-phenylenevinylene) (**P3**)

^{1}H NMR ($CDCl_3$): $\delta = 7.75$–6.92 (arom., *trans*-vinylene) 6.72 (dd, terminal vinyl (α-H)), 5.79 (d, terminal vinyl (*cis*-β-H)), 5.5 (br m, vinylidene), 5.28 (d, terminal vinyl (*trans*-β-H)), 4.57, 4.54 (Ar-CH_2-O), 3.63–3.45 (-OCH_2-), 1.77–1.15 (methylene), 0.89 (t, methyl) ppm; ^{13}C NMR ($CDCl_3$): $\delta = 139.7$, 138.1, 136.9, 128.7, 127.2, 125.2, 124.6, 123.7, 114.1 (arom., vinylene, terminal vinyl), 72.9, 71.0 (-CH_2-O-CH_2-), 32.2–22.4 (methylene), 14.4 (methyl) ppm; FT IR (NaCl): 3073, 3021 (CH stretch, vinylene, vinyl, arom.), 2922, 2847 (CH stretch, aliph.), 1634, 1593, 1521 (C=C stretch, vinylene + arom.), 1457, 1361, 1104 (C–O stretch), 960 (CH out of plane, *trans*-vinylene), 906 (CH out-of-plane, terminal vinyl), 855, 813 (CH out-of-plane, arom.), 735 (CH_2 rocking), 695 (ring deformation, arom.) cm^{-1}.

Oligo-(4,4′-biphenylenevinylene-co-4-decyloxy-1,3-phenylenevinylene) (**P4**)

^{1}H NMR ($CDCl_3$): $\delta = 7.85$–6.95 (arom., *trans*-vinylene), 6.85 (dd, terminal vinyl (α-H)), 5.81 (d, terminal vinyl (*cis*-β-H)), 5.29 (terminal vinyl (*trans*-β-H)), 4.22–3.88 (-OCH_2-), 2.02–1.12 (methylene), 1.05–0.72 (methyl) ppm; FT IR (NaCl): 3018 (CH stretch, vinylene, vinyl, arom.), 2955, 2926, 2852 (CH stretch, aliph.), 1630, 1601, 1510 (C=C stretch, vinylene + arom.), 1475, 1359, 1252, 1105 (C–O stretch), 1026, 970 (CH out-of-plane, *trans*-vinylene), 907 (CH out-of-plane, terminal vinyl), 817 (CH out-of-plane, arom.), 744, 699 (ring deformation, arom.) cm^{-1}.

Acknowledgements

We thank the *SFB Electroactive Materials* for financial support.

References

[1] Gustavson G, Cao Y, Treacy GM, Klavetter F, Colaneri N, Heeger AJ (1992) Nature **357**: 477
[2] Berggren M, Granström M, Inganäs O, Andersson M (1995) Adv Mater **7**: 900
[3] Greenham NC, Moratti SCV, Bradley DDC, Friend RH, Holmes AB (1993) Nature **365**: 628
[4] Burroughes JH, Bradley DDC, Brown AR, Marks RN, MacKay K, Friend RH, Burns PL, Holmes AB (1990) Nature **347**: 539
[5] Winkler B, Tasch S, Leising G, Stelzer F (1997) Synth Met **84**(1–3): 383
[6] Mullner R, Winkler B, Stelzer F, Tasch S, Hochfilzer C, Leising G (1999) Synth Met **108**(2): 129
[7] Winkler B, Meghdadi F, Tasch S, Mullner R, Resel R, Saf R, Leising G, Stelzer F (1998) Optical Materials **9**: 159
[8] Sariciftci NS, Smilowitz L, Heeger AJ, Wudl F (1992) Science **258**: 1474
[9] Friend RH, Greenham NC (1998) In: Skotheim TA, Elsenbaumer RL, Reynolds JR (eds) Handbook of Conducting Polymers, 2nd edition. Dekker, New York Basel Hong Kong, p 823
[10] Greenham NC, Friend RH (1995) Solid State Physics **49**: 32
[11] Kalinowski J (1993) Proc SPl **1910**: 135
[12] Kraft A, Grimsdale AC, Holmes AB (1998) Angew Chem **110**: 416
[13] Segura JL (1998) Acta Polym **49**: 319
[14] Vaubel G, Bässler H, Möbius D (1991) Chem Phys Lett **10**: 334
[15] Brown AR, Bradley DDC, Burn PL, Burroughes JH, Friend RH, Greenham N, Holmes AB, Kraft A (1992) Appl Phys Lett **61**: 2793
[16] Yan M, Rothberg LJ, Papadimitrakopoulos F, Galvin ME, Miller TM (1994) Phys Rev Lett **73**: 744
[17] Köpping-Grem G, Leising G, Schimetta M, Stelzer F, Huber A (1996) Synth Met **76**: 53
[18] Braun D, Staring EGJ, Demandt RCJE, Rikken GLJ, Kessener YARR, and Verhuizen AHJ (1994) Synth Met **66**: 75

[19] Friend RH, Denton GM, Halls JJM, Harrison NT, Holmes AB, Köhler A, Lux A, Moratti SC, Pichler K, Tessler N, Towns K (1997) Synth Met **84**: 463
[20] Carter J, Grizzi I, Heeks SK, Lacey DJ, Latham SG, May PG, Ruiz de los Panos O, Pichler K, Towns C, Wittmann HF (1997) Appl Phys Lett **71**: 34
[21] O'Brien D, Bleyer A, Lidzey DG, Bradley DDC, Tsutsui T (1997) J Appl Phys **82**(5): 2662
[22] Pang Y, Li J, Hu B, Karasz FE (1999) Macromolecules **32**: 3946
[23] Ahn T, Jang MS, Shim HK, Hwang DH, Zyung T (1999) Macromolecules **32**: 3279
[24] Lee JK, Schrock RR, Baigent DR, Friend RH (1995) Macromolecules **28**: 1966
[25] Hesenmann P, Vestweber H, Pommerehne J, Mahrt RF, Greiner A (1995) Adv Mater **7**(4): 388
[26] Hochfilzer C, Tasch S, Winkler B, Huslage J, Leising G (1997) Synth Met **85**: 1271
[27] Jenekhe SA, Osaheni SA (1994) Science **265**: 765
[28] Vestweber H, Oberski J, Greiner A, Heitz W, Mahrt RF, Bässler H (1993) Adv Mater Opt Electr **2**: 197
[29] Winkler B, Ungerank M, Stelzer F (1996) Macromol Chem & Phys **197**: 2343
[30] Ungerank M, Winkler B, Eder E, Stelzer F (1997) Macromol Chem & Phys **198**: 1391
[31] Winkler B, Rehab A, Ungerank M, Stelzer F (1997) Macromol Chem & Phys **198**: 1417
[32] Winkler B, Tasch S, Zojer E, Ungerank M, Leising G, Stelzer F (1996) Synth Met **83**(3): 177
[33] Mullner R, Noirez L, Zojer E, Stelzer F, Leising G (1999) MRS fall 99 proceedings, session BB, paper 5
[34] Schlick H, Stelzer F, Tasch S, Leising G, J Mol Catal A: Chemical (2000) **160**: 71
[35] Heck RF (1981) Org React **27**: 345
[36] Greiner A, Heitz W (1996) In: Salamone J (ed) The Polymeric Encyclopedia, vol 7. CRC Press, Boca Raton, p 4865
[37] Klingelhöfer S, Schellenberg C, Pommerlehne J, Bässler H, Greiner A, Heitz W (1997) Macromol Chem & Phys **198**(5): 1511
[38] Hilberer A, Brouwer JB, van der Scheer B-J, Wildeman J, Hadziioannou G (1995) Macromolecules **28**: 4525
[39] Schlick H, Stelzer F, Meghdadi F, Leising G (in preparation); Schlick H, PhD Thesis, TU Graz, 2000
[40] Jacobs S, Eevers W, Verreyt G, Geise HJ (1993) Synth Met **61**: 189
[41] Pouchert CJ, Behnke J (1993) The Aldrich Library of ^{13}C and ^{1}H FT NMR Spectra, vol 2, p 37B
[42] Bredas JL (1986) In: Skotheim TA (ed) Handbook of Conducting Polymers, vol 2. Dekker, New York Basel Hong Kong, p 859
[43] Le Bigeot Y, Delmas M, Gaset A (1982) Synth Comm **12**: 107
[44] Thorn-Csanyi E, Kraxner P (1997) Macromol Chem Phys **198**: 3827
[45] Becker HGO, Berger W, Domschke G, Fanghänel E, Faust J, Fischer M, Gentz F, Gewald K, Gluch R, Mayer R, Müller K, Pavel D, Schmidt H, Schollberg K, Schwetlick K, Seiler E, Zeppenfield G (eds) (1986) Organikum, 19. Auflage. Johann Ambrosius Barth, Leipzig Berlin Heidelberg
[46] Henze O, Lehmann U, Schlueter AD (1999) Synthesis **4**: 683

Received May 30, 2000. Accepted (revised) December 19, 2000

Donor-Acceptor Substituted Cyclohexasilanes: Materials with Potential Nonlinear Optical Properties

Christa Grogger, Hermann Rautz, and **Harald Stüger***

Institut für Anorganische Chemie, Technische Universität Graz, A-8010 Graz, Österreich

Summary. The dipolar cyclohexasilanes 1-ferrocenyl-4-((2,2-dicyanoethenyl)-phenyl)-decamethylcyclohexasilane (**1**) and 1-ferrocenyl-3-((2,2-dicyanoethenyl)-phenyl)-decamethylcyclohexasilane (**2**) were synthesized by conventional methods and fully characterized. Electronic absorption spectroscopic studies suggest increased σ-π interactions in the cyclohexasilane derivatives when compared to their open-chained analogues. Small but noticeable electron transmission effects *via* the cyclohexasilanyl group can be derived from cyclic voltammetric data. Significantly increased values of the quadratic hyperpolarizabilities β in solution for **1** and **2** are shown by preliminary results of measurements using the EFISH technique.

Keywords. Polysilanes; UV spectroscopy; Cyclic voltammetry; Nonlinear optics.

Introduction

Interactions in materials showing nonlinear optical (NLO) properties change the nature of the incident light such that new field components are produced. These materials are of technological importance in areas that utilize optical devices, with potential applications in optical signal processing, switching, and frequency generation [1]. Whereas inorganic solids such as $LiNbO_3$ and KH_2PO_4 have traditionally been the NLO materials of choice, recent results suggest that molecule-based organic or organometallic materials possess superior NLO characteristics [2]. Key advantages of all organic NLO materials include the high intrinsic nonlinearities of individual organic molecules, the ability to use molecular engineering to tailor properties for specific applications, low DC dielectric constants, low temperature processing, and sometimes very fast response times. Organometallic structures additionally permit NLO responses to be tuned in ways not possible for purely organic molecules simply by variation in metal, oxidation state, ligand environment, and geometry. The search for novel organic and organometallic molecules displaying optimal nonlinear optical properties is therefore currently an intense area of research in both academia and industry.

* Corresponding author

Quite recently, organic materials with second-order nonlinear optical properties have been the subject of intense investigation, leading to the development of certain structure/NLO property relationships. Thus, large second-order molecular hyperpolarizabilities (β) are always associated with chromophores comprised of electron donors and acceptors linked by a conjugated π-system. The majority of systems which have been explored for second order nonlinear optics fall into the classes of donor-acceptor substituted benzenes, biphenyls, stilbenes, thiophenes, and related structures. The nonlinearity of such molecules can be enhanced by either increasing the conjugation length, thereby increasing electron delocalization, or increasing the strength of donor or acceptor groups to improve electron asymmetry within the molecule. Several reports have appeared in the recent literature quoting appreciable second-order nonlinear optical susceptibilities for organometallic chromophores containing ferrocene as a donor. Two molecules, *trans*-1-ferrocenyl-2-(N-methylpyridinium-4-yl)-ethylene iodide [3] and *cis*-1-ferrocenyl-2-(4-nitrophenyl)-ethylene [4] have been reported to possess rather large powder nonlinearities (220 and 62×urea, respectively) and have sparked subsequent research activity in this area. Nowadays, ferrocene-based systems are prominent in the class of organometallic NLO compounds [2b, c], what is mainly due to attractive features like stability towards air, elevated temperatures, or light and synthetic versatility.

An undesirable feature associated with nearly all organic and organometallic NLO materials is their often strong absorption in the visible light region, thus dramatically limiting the applicability of the corresponding materials *e.g.* for frequency doubling of the fundamental wavelength of a 820 nm laser requiring transparency at the first and second harmonic wavelengths near 800 and 400 nm. Consequently, the search for new compounds with increased transparency combined with high nonlinear efficiency remains a crucial challenge for various applications. Recently, several reports have appeared in the literature [5] concerning the synthesis and characterization of dipolar silicon compounds of the type shown below with an extended transparency range combined with interesting nonlinear optical activities.

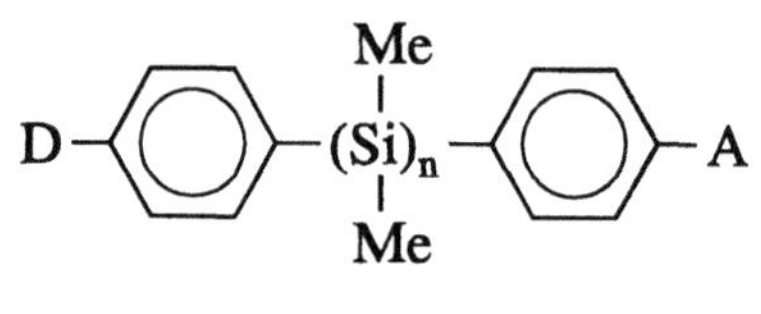

I

$n = 2, 3, 4, 6$

$D = NMe_2, MeS, MeO, F, SiMe_3, Si_2Me_5$

$A = HC{=}C(CN)_2, CF_3, SO_2R$

Connecting the donor and acceptor groups by polysilane backbones is considered to be attractive because of the following features: *i*) increased polarizability of silicon chains as compared to the carbon analogues, *ii*) σ-π-interactions (σ(Si)-π(C)), and *iii*) electron delocalization over the Si–Si bonds *via* σ-conjugation [6].

Silicon, however, was found to be only a weak charge transmitter when donor and acceptor phenyl-Si-phenyl moieties are connected in compounds of type **I**. The nonlinear optical behaviour as shown by measurements of the quadratic hyperpolarizablities β using the EFISH technique easily can be accounted for by a vectorial additive model where the charge transfer interactions contributing to the overall nonlinearity are confined within the donor and the acceptor subunits. Quite similar results have been independently obtained by *Pannell et al.* [7] and in our laboratories [8] for linear polysilanes containing ferrocene as a donor and various organic acceptor groups.

In the present work the influence of the structure of the polysilane bridge on charge transfer properties in related donor-acceptor substituted organosilicon compounds is studied. In cyclic polysilanes, delocalization of σ(Si-Si) electrons over the ring silicon atoms gives these molecules a variety of properties resembling those of aromatic hydrocarbons [9]. Cyclopolysilanes, therefore, might be much more effective charge transmitters in dipolar structures like **I** compared to the open chained analogues. Thus, we report the synthesis and UV/Vis absorption spectroscopic as well as electrochemical studies of the novel cyclohexasilane derivatives **1** and **2** bearing ferrocene as a donor and the (2,2-dicyanoethenyl)-phenyl group as an acceptor.

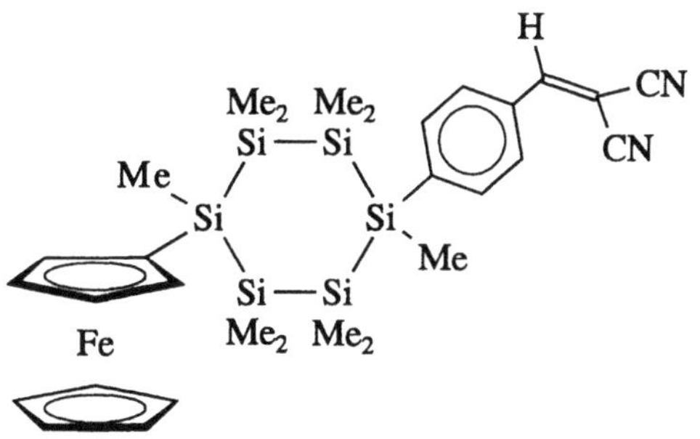

1

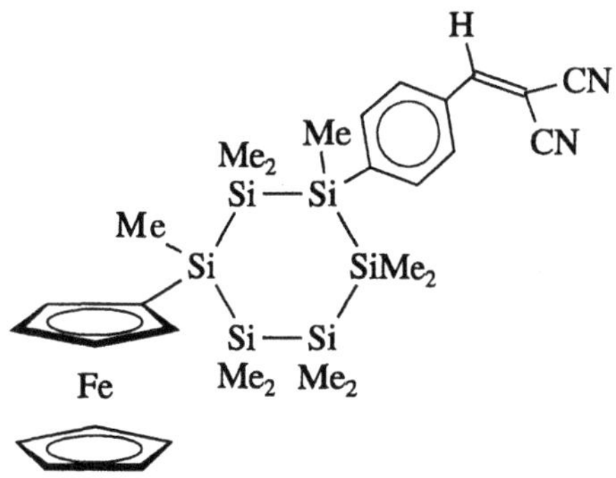

2

Results and Discussion

Synthesis

The method used to prepare compounds **1** and **2** in this work mainly follows reaction sequences originally developed by *Pannell* [7] and *Mignani* [5b] for the synthesis of similar donor-acceptor substituted oligosilanes. The description of the synthesis of **1** is shown in Scheme 1. Compound **2** has been prepared analogously starting from 1,3-dichlorodecamethylcyclohexasilane. In both cases, the *E,E*-isomer has been isolated selectively by crystallization from hexane and used for further studies.

Closely related reaction sequences were used to prepare the reference compounds ((2,2-dicyanoethenyl)-phenyl)-undecamethylcyclohexasilane (**3**) and ferro-

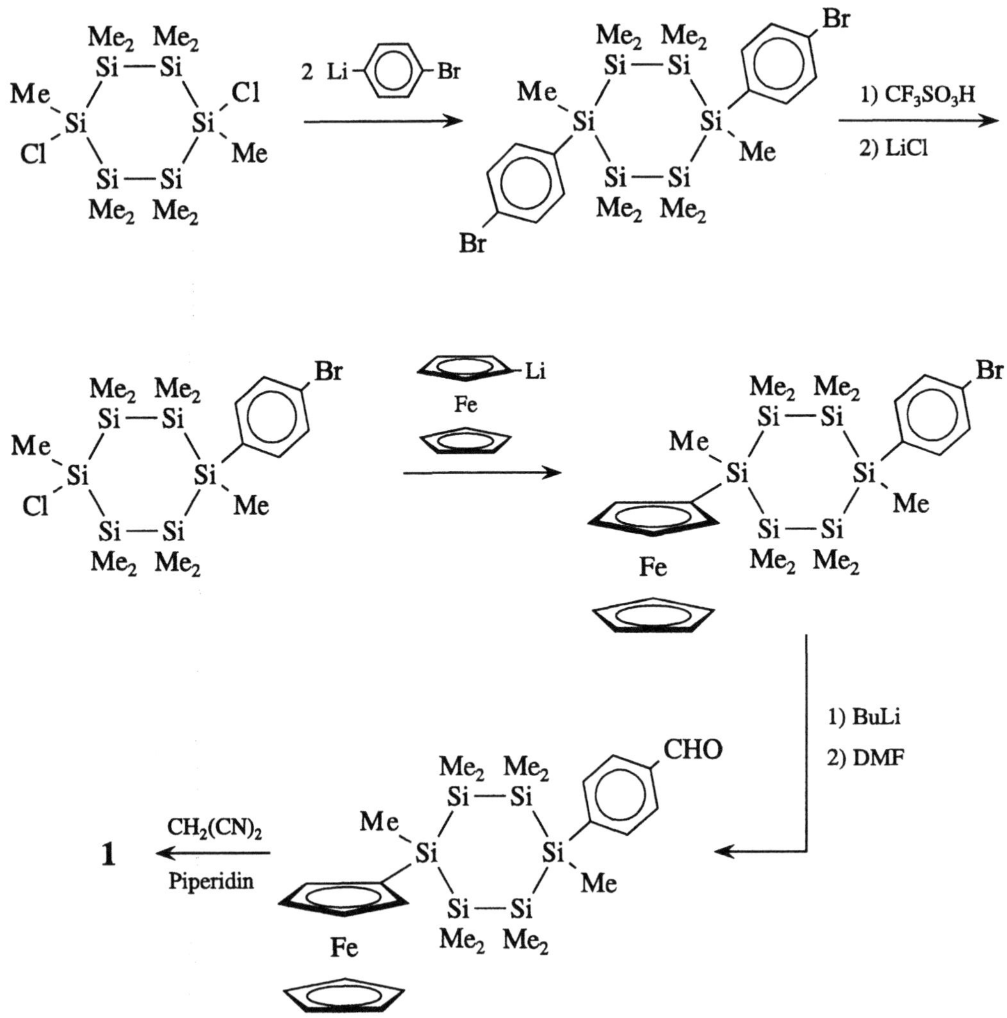

Scheme 1. Synthesis of compound **1**

cenylundecamethylcyclohexasilane (**4**) from chloroundecamethylcyclohexasilane and 1,4-dibromobenzene or ferrocenyllithium, respectively (Eqs. (1) and (2)).

$$Me_{11}Si_6Cl \xrightarrow[3)\ DMF]{1)\ LiPhBr\ 2)\ BuLi} Me_{11}Si_6PhCHO \xrightarrow[Piperidine]{CH_2(CN)_2} \underset{\mathbf{3}}{Me_{11}Si_6PhCH{=}C(CN)_2} \quad (1)$$

$$Me_{11}Si_6Cl \xrightarrow{ferrocenyllithium} \underset{\mathbf{4}}{Me_{11}Si_6((C_5H_4)Fe(C_5H_5))} \quad (2)$$

All products were characterized by standard spectroscopic techniques (^{29}Si, ^{13}C, and ^{1}H NMR, EI-MS) and elemental analysis. The results are in accordance with the proposed structures in all cases.

Table 1. Electronic absorption and cyclic voltammetric data for compounds **1–4**

	λ_{max}/nm^{a}	$\varepsilon/mol^{-1}\cdot cm^{-1\,a}$	$E_f/V^{b,c}$	E_{pa}/V^{b}	E_{pc}/V^{b}
1	230 sh	20000			
	279	10100	0.04	0.97	−1.68
	368	17500			
2	230 sh	21000			
	281	12600	0.04	0.97	−1.67
	367	17300			
3	289	12300		1.10	−1.69
	366	23000			
4	456[d]	220[d]			
	328[d]	160[d]	0.02	0.93	
	260 sh	8500			
	225 sh	20300			

[a] Data measured for cyclohexane solutions of $c = 2\cdot 10^{-5}\,M$; [b] conditions: sample concentration $10^{-3}\,M$, electrolyte Bu_4NBF_4 (0.1 *M*), CH_2Cl_2 solution, scan rate 50 mV, potentials quoted *vs.* ferrocene/ferrocenium; [c] $E_f = (E_{pa} + E_{pc})/2$, $\Delta E_p = 180$ mV; [d] sample concentration $4\cdot 10^{-4}\,M$

Electrochemical studies

To gain more insight into the electronic properties of compounds **1** and **2**, we studied their redox behavior by cyclic voltammetry. The pertinent data are presented in Table 1 together with data obtained for the reference compounds **3** and **4** which represent reasonable model systems to estimate the properties of the donor and the acceptor group in the absence of interaction. The representative cyclic voltammogram of compound **1** is shown in Fig. 1.

The irreversible wave at high cathodic potential can be attributed to reduction processes of the dicyanovinyl group. The electron donating ferrocene moiety affects the reduction potentials; thus, **1** and **2** exhibit 10 and 20 mV anodic shifts, respectively, in the reduction potential relative to **4**.

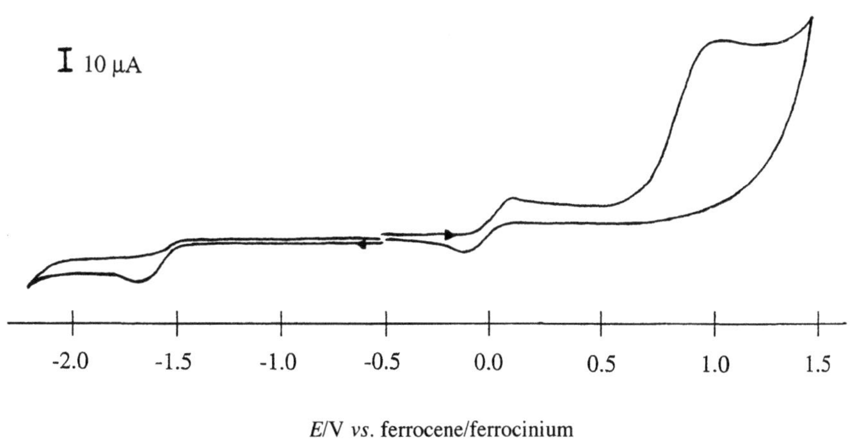

Fig. 1. Cyclic voltammogram for **1**

The first anodic wave observed in **1**, **2**, and **4** displays electrochemical reversibility with a peak potential E_{pa} positioned close to that of the unsubstituted ferrocene. It is therefore reasonable to assign it to the ferrocene moiety. Due to the electron withdrawing dicyanovinyl group, **1** and **2** exhibit 20 mV anodic shifts relative to **4**. The ΔE_p values $(E_{p(anodic)} - E_{p(cathodic)})$ of the standard ferrocene and the compounds used in this study fall in the same range. Thus, the ferrocene oxidation may be termed as reversible or *quasi*-reversible one electron transfer. In contrast, the second oxidation step in **1**, **2**, and **4** (and the first one in **3**) displays complete electrochemical irreversibility. Comparing the shape and maximum with unsubstituted Si_6Me_{12} $(E_{pa} = 1.06\,V)$ it seems clear that this irreversible oxidation occurs in the cyclohexasilanyl part of the molecule. Including reference compounds **3** and **4**, the differences in the oxidation potentials reflect the electron withdrawing or donating properties of the substituents. Thus, the cyclohexasilanyl group shows significant higher oxidation potentials in **1** and **2** than in **3**, but lower ones than in **4**.

Although a measuring accuracy below 10 mV seems not defendable with regard to the rather high peak separation of 180 mV for the ferrocene/ferrocenium couple, the data underline that there is a noticeable transmission effect *via* the cyclohexasilanyl group. However, it is rather small compared to that of a vinyl bridge. Cyclic voltammetric studies of 2-(4-(2-ferrocenylvinyl)-benzylidene)-malononitrile showed an anodic shift of 70 mV for the ferrocene group [10], which is about twice as much compared to **1** and **2**.

UV/Vis spectra

Solution electronic absorption spectroscopic studies of compounds designed to possess second order NLO properties are important to determine the transparency region and to study the solvatochromic behavior of the sample. Pronounced solvatochromism is generally considered as indicative of intramolecular charge transfer upon electronic excitation, what is a prerequisite for high molecular hyperpolarizabilities β. The electronic absorption spectra of compounds **1** and **2** were recorded in solvents of different polarity; the corresponding data are presented in Tables 1 and 2. The UV absorption spectra of **1** and **2** are displayed in Fig. 2. Figure 3 shows the spectra of the reference chromophores **3** and **4**.

The spectrum of compound **4** exhibits two weak absorption maxima near 460 and 330 nm (compare Fig. 2) typical for silylated ferrocenes, which are nearly unaffected by the structure of the silyl substituent attached [11]. These bands apparently are superimposed by much stronger bands appearing around 280 and 370 nm in the spectra of **1**, **2**, and **3**.

As shown in Fig. 2, the spectra of **1** and **2** resemble the calculated sum spectrum of the separate chromophores as represented by compounds **3** and **4**. The absorption spectra of **1** and **2**, therefore, seem to fulfill the classical expectation for nonconjugatively connected chromophores.

Furthermore, the solvatochromic shifts are small (< 5 nm) when the solvent is changed from apolar to polar (compare Table 2). In agreement with the electrochemical studies, the above features indicate rather limited intramolecular charge

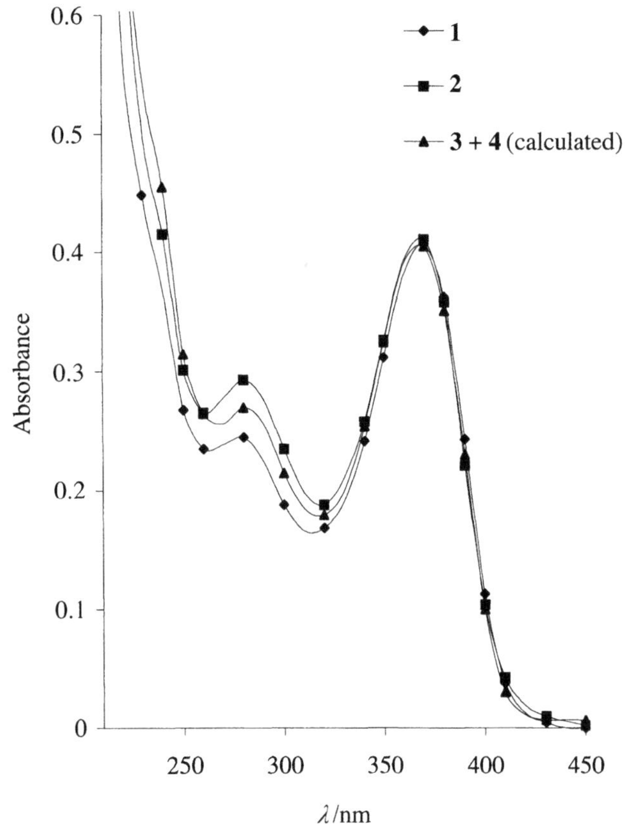

Fig. 2. Electronic absorption spectra of **1** and **2** ($c = 2 \cdot 10^{-5} M$) in cyclohexane solution and calculated sum spectrum of **3** and **4**

transfer from the donor to the acceptor group *via* the polysilanyl bridge upon electronic excitation in **1** and **2**.

Most striking is the close similarity of the absorption spectra of **1**, **2**, and **3**, regardless of the presence of ferrocene and of the relative position of the donor and the acceptor moiety at the cyclohexasilane ring in **1** and **2**. The 280 nm band appearing in the spectra of **1**, **2**, and **3** can be assigned to the perturbed σ-σ^* transition within the cyclohexasilanyl ring simply by comparison with literature values [12]. According to literature data available for linear polysilanes containing the dicyanovinyl acceptor along with various donor groups [5a,b,f], the 370 nm band can be assigned to the π-π^* transition of the acceptor-containing phenyl ring. In **1**, **2**, and **3**, however, the band appears considerably red shifted (20–30 nm) due to enhanced σ-π interaction of the cyclohexasilane σ-(Si-Si)- and the phenyl π-system. Compared to open chained organosilicon fragments, therefore, the cyclohexasilane ring should be markedly more effective as an electron releasing group [13]. In preliminary studies quite exceptional values of the first hyperpolarizability β have been measured for compounds **1** and **2** (up to 3 times that of the open chained analogues) using the EFISH technique [14]. This improved optical nonlinearity very likely may be attributed to the increased donor capacity of the cyclohexasilane ring mentioned above, hence facilitating intramolecular charge

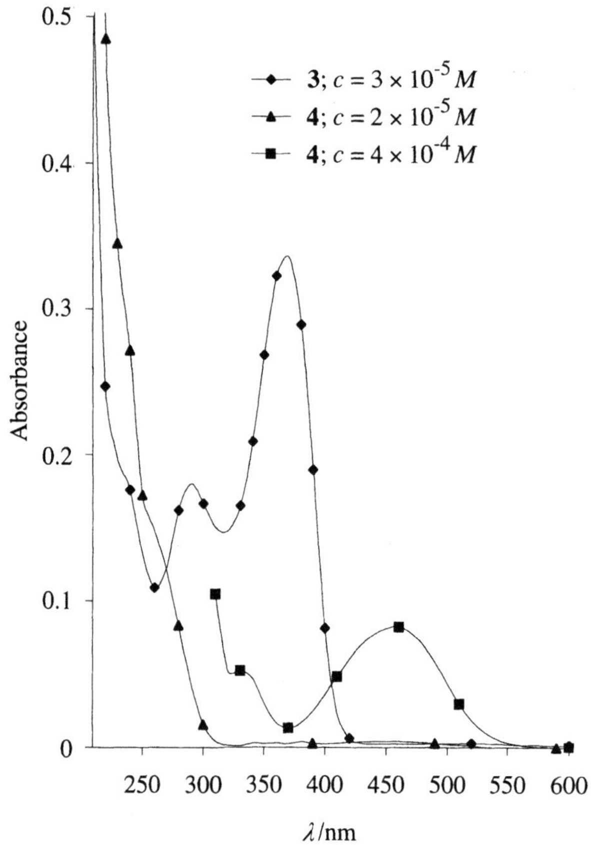

Fig. 3. Electronic absorption spectra of **3** and **4** in cyclohexane solution

Table 2. Solvatochromic data for compounds **1–4**

	C_6H_{12}	CH_3CN	CH_2Cl_2
1[a]	279	281	278
	368	364	363
2[a]	281	280	279
	367	363	364
3[a]	289	292	296
	366	364	378
4[b]	328	325	325
	456	455	458

[a] Data measured for solutions of $c = 2 \cdot 10^{-5}\,M$; [b] sample concentration $10^{-3}\,M$

transfer from the Si_6 fragment to the dicyanovinyl acceptor *via* the aromatic ring [5a]. The question to what extent the presence of the ferrocene group is crucial for the magnitude of β in **1** and **2** at all is currently the matter of ongoing investigations.

Conclusions

In conclusion, it has been shown by UV/Vis and electrochemical studies that the ferrocenyl donor and the (2,2-dicyanoethenyl)-phenyl acceptor group interact only weakly when linked through a cyclohexasilanyl bridge. UV/Vis absorption spectroscopy, however, provides evidence that σ(Si-Si)-π(phenyl) interactions and, consequently, the electron releasing character of the polysilanyl fragment are enhanced in the cyclohexasilane derivatives investigated as compared to their open-chained analogues. Improved optical nonlinearities observed for compounds **1** and **2** in preliminary studies may be due to more facile charge transfer processes within the cyclohexasilanyl-phenyl acceptor moiety. Further studies to substantiate this assumption are currently in progress in our laboratories.

Experimental

All experiments were performed under a nitrogen atmosphere using standard *Schlenk* techniques. All solvents were dried with sodium-potassium alloy and distilled under nitrogen prior to use. Malononitrile, triflic acid, and 1,4-dibromobenzene were used as purchased without further purification. N,N-Dimethylformamide was allowed to stand over 4 Å molecular sieve and distilled from CaH_2 prior to use. Ferrocenyllithium [15] and the starting silicon compounds 1-chloroundecamethyl-cyclohexasilane [16], 1,3-dichlorodecamethylcyclohexasilane, and 1,4-dichlorodecamethylcyclohexasilane [17, 18] were synthesized as previously reported. Cyclic voltammetric studies were performed at room temperature with a Wenking LB 95 M potentiostat and a Wenking POS 73 scan generator using a 0.1 *M* solution of Bu_4NBF_4 in CH_2Cl_2, a platinum disk (500 µm) microelectrode, a Pt counter electrode, and a Ag/AgCl reference electrode. The sample concentration was 10^{-3} *M* and the scan rate 50 mV. Potentials are quoted relative to the ferrocene/ferrocinium redox couple [19].

1,4-Bis-(4-bromophenyl)-decamethylcyclohexasilane

To a solution of 35.0 g (89.8 mmol) of 1,4-dichlorodecamethylcyclohexasilane in 100 cm^3 diethyl ether, a -70°C cold solution of 4-bromophenyllithium (prepared from 65.1 g, 276.1 mmol of 1,4-dibromobenzene, and 276.1 mmol of *n*-butyllithium in 2000 cm^3 of diethyl ether) was added dropwise [20]. The mixture was stirred for 2 h at room temperature and subsequently refluxed for 1 h. After aqueous workup with 50 cm^3 of saturated NH_4Cl solution, the combined organic layers were dried over Na_2SO_4, and the solvent was stripped off in vacuum. Recrystallization of the crude product from 200 cm^3 of 1-propanol affords pure white crystals of 1,4-*bis*-(4-bromophenyl)-decamethylcyclohexasilane in 65% yield.

1-Chloro-4-(4-bromophenyl)-decamethylcyclohexasilane

5.9 cm^3 (67.6 mmol) of triflic acid were added to a solution of 40.6 g (64.4 mmol) of 1,4-*bis*-(4-bromophenyl)-decamethylcyclohexasilane in 1000 cm^3 of toluene at -20°C over a period of 7 h. The mixture was stirred at room temperature overnight, and 300 cm^3 of toluene were removed in vacuum and replaced by diethyl ether. 6.1 g (144.2 mmol) of anhydrous LiCl were added at 0°C, and the mixture was stirred for another 12 h at room temperature. After removal of the solvent in vacuum, 200 cm^3 of petroleum ether were added, and the solution was filtered. Petroleum ether was stripped off from the filtrate, and the resulting solid residue was sublimed at 120°C and 10^{-2} mbar to give the title compound in 75% yield.

1-Ferrocenyl-4-(4-bromophenyl)-decamethylcyclohexasilane

A solution of ferrocenyllithium (prepared from 21 g (113 mmol) of ferrocene and 75 mmol of *tert*-butyllithium in 150 cm^3 of *THF*; ferrocenyllithium was crystallized by cooling the solution to −20°C overnight, filtering and washing with cold *THF* in order to remove excess ferrocene and *tert*-butyllithium) was added slowly to a stirred solution of 24.9 g (48.7 mmol) of 1-chloro-4-(4-bromophenyl)-decamethylcyclohexasilane in 50 cm^3 of *THF* at room temperature. The mixture was refluxed overnight. Subsequently, the solvent was stripped off, and residual ferrocene was removed by sublimation at 50°C and 10^{-2} mbar. After aqueous workup of the resulting orange residue dissolved in diethyl ether with saturated NH_4Cl solution, the combined organic layers were dried over Na_2SO_4 and concentrated. Chromatography on a silica gel column developed by a mixture of heptane/toluene (95:5) affords the desired compound in 27% yield.

1-Ferrocenyl-4-(4-formylphenyl)-decamethylcyclohexasilane

To a solution of 8.62 g (13.1 mmol) of 1-ferrocenyl-4-(4-bromophenyl)-decamethylcyclohexasilane in 80 cm^3 of *THF* at −78°C, a pentane solution of 26.1 mmol of *tert*-butyllithium was added slowly. The mixture was stirred for another 5 min and 10.4 cm^3 of dry *DMF* were added. After stirring the mixture overnight at room temperature, aqueous workup with saturated NH_4Cl solution and recrystallization of the crude product from pentane afforded 1-ferrocenyl-4-(4-formylphenyl)-decamethylcyclohexasilane in 55% yield.

1-Ferrocenyl-4-((2,2-dicyanoethenyl)-phenyl)-decamethylcyclohexasilane (**1**; *E,E*-isomer)

To a solution of 4.3 g (7.1 mmol) of 1-ferrocenyl-4-(4-formylphenyl)-decamethylcyclohexasilane dissolved in a mixture of 15 cm^3 of *THF* and 80 cm^3 of deoxygenated EtOH, 0.52 g (7.8 mmol) of solid malononitrile and a drop of piperidine were added. An immediate colour change from orange to dark red was observed. The mixture was stirred for 45 min (DC monitoring recommended), and the solvents were removed in vacuum. After column chromatography (silica gel) with toluene as the mobile phase and subsequent recrystallization from hexane, exclusively the *E,E*-isomer of the title compound was obtained in 50% yield.

M.p.: 152–153°C; $C_{30}H_{44}FeN_2Si_6$ (657.06); found: C 54.94, H 6.81, N 4.29; calcd.: C 54.84, H 6.75, N 4.21; ^{29}Si NMR (Toluol/D_2O, ext. *TMS*, 59.62 MHz): $\delta = -39.94$, −40.68, −41.00, −41.64 ppm; ^{1}H NMR (C_6D_6, ext. *TMS*, 300.13 MHz): $\delta = 7.354$ (s, 2H, phenyl), 7.151 (s, 2H, phenyl), 6.558 (s, 1H, HC=), 4.21 (t, 2H, ferrocene), 4.11 (s, 5H, ferrocene), 4.01 (t, 2H, ferrocene), 0.517 (s, 3H, $Si(CH_3)$), 0.401 (s, 3H, $Si(CH_3)$), 0.328 (s, 6H, $Si(CH_3)$), 0.273 (s, 6H, $Si(CH_3)$), 0.261 (s, 6H, $Si(CH_3)$), 0.116 (s, 6H, $Si(CH_3)$) ppm; ^{13}C NMR (C_6D_6, ext. *TMS*, 75.47 MHz): $\delta = 159.09$ (=*C*(CN)$_2$), 149.15, 135.73, 131.16, 129.85 (phenyl), 114.60, 113.62 (=C(*C*N)$_2$), 82.80 (HC=), 73.99, 71.35, 69.00 (ferrocene), −3.57, −3.79, −4.62, −5.24, −6.20, −6.42 ($Si(CH_3)$) ppm; MS (70 eV): m/z (%) = 656 (2.6, M^+).

1-Ferrocenyl-3-((2,2-dicyanoethenyl)-phenyl)-decamethylcyclohexasilane (**2**; *E,E*-isomer)

The procedure followed was that used for **1** with 35.0 g (89.8 mmol) of 1,3-dichlorodecamethylcyclohexasilane, 65.1 g (276.1 mmol) of 1,4-dibromobenzene, and 276.1 mmol of *n*-butyllithium. The *E,E*-isomer was isolated by fractional crystallization from hexane in yields comparable to **1**.

M.p.: 152°C; $C_{30}H_{44}FeN_2Si_6$ (657.06); found: C 55.20, H 6.82, N 4.28; calcd.: C 54.84, H 6.75, N 4.21; ^{29}Si NMR (Toluol/D_2O, ext. *TMS*, 59.62 MHz): $\delta = -38.93$, −40.99, −41.08, −41.19, −41.81 ppm; ^{1}H NMR (C_6D_6, ext. *TMS*, 300.13 MHz): $\delta = 7.30$, 7.28, 7.25, 7.23 (4H, phenyl), 6.51 (s, 1H, HC=), 4.30–4.26 (m, 2H, ferrocene), 4.09 (s, 5H, ferrocene), 4.05–3.94 (m, 2H, ferrocene), 0.53 (s, 3H, $Si(CH_3)$), 0.45 (s, 3H, $Si(CH_3)$), 0.35 (s, 3H, $Si(CH_3)$), 0.34 (s, 3H, $Si(CH_3)$), 0.27 (s,

3H, Si(CH_3)), 0.22 (s, 3H, Si(CH_3)), 0.21 (s, 3H, Si(CH_3)), 0.19 (s, 6H, Si(CH_3)), 0.16 (s, 3H, Si(CH_3)) ppm; ^{13}C NMR (C_6D_6, ext. *TMS*, 75.47 MHz): $\delta = 159.13$ (=*C*(CN)$_2$), 148.94, 135.93, 131.08, 129.81 (phenyl), 114.62, 113.61 (=C(*C*N)$_2$), 82.66 (HC=), 74.61, 73.93, 71.64, 71.38, 70.58, 69.06, 69.00 (ferrocene), −3.43, −3.64, −3.32, −4.79, −4.85, −4.87, −5.76, −5.80, −6.78 (Si(CH_3)) ppm; MS (70 eV): m/z (%) = 656 (100, M^+).

(4-Bromophenyl)-undecamethylcyclohexasilane

The procedure followed was that used for 1,4-*bis*-(4-bromophenyl)-decamethylcyclohexasilane with 4.0 g (10.8 mmol) of chloroundecamethylcyclohexasilane, 3.7 g (16.2 mmol) of 1,4-dibromobenzene, and 16.2 mmol of *n*-butyllithium. Yield: 5 g (90% based on chloroundecamethylcyclohexasilane).

(4-Formylphenyl)-undecamethylcyclohexasilane

The procedure followed was that used for 1-ferrocenyl-4-(4-formylphenyl)-decamethylcyclohexasilane with 4.3 g (8.8 mmol) of (4-bromophenyl)-undecamethylcyclohexasilane in 50 cm^3 of *THF*, 17.6 mmol of *tert*-butyllithium, and 7 cm^3 of *DMF*. Yield: 3.5 g (91%).

(2,2-Dicyanoethenyl)-phenylundecamethylcyclohexasilane (**3**)

The procedure followed was that used for 1-ferrocenyl-3-((2,2-dicyanoethenyl)-phenyl)-decamethylcyclohexasilane with 3.1 g (7.1 mmol) of (4-formylphenyl)-undecamethylcyclohexasilane in 15 cm^3 of *THF* and 80 cm^3 of deoxygenated ethanol and 0.5 g (7.7 mmol) of solid molononitrile.

Yield: 2.8 g (81%); m.p.: 77–78°C; $C_{21}H_{38}N_2Si_6$ (487.29); found: C 51.85, H 8.09; calcd.: C 51.78, H 7.86; ^{29}Si NMR (Toluol/D_2O, ext. *TMS*, 59.62 MHz): $\delta = -39.50$, −41.08, −41.47, −41.98 ppm; ^{1}H NMR (C_6D_6, ext. *TMS*, 300.13 MHz): $\delta = 7.0$–7.4 (m, 4H, phenyl), 6.65 (s, 1H, HC=), 0.39 (s, 3H, Si(CH_3)), 0.22 (s), 0.213 (s), 0.210 (s), 0.20 (s, 24H, Si(CH_3)), 0.11 (s, 3H, Si(CH_3)) ppm; ^{13}C NMR (C_6D_6, ext. *TMS*, 75.47 MHz): $\delta = 159.74$ (=*C*(CN)$_2$), 148.70, 135.21, 130.53, 129.29 (phenyl), 114.07, 113.05 (=C(*C*N)$_2$), 82.03 (HC=), −4.68, −5.43, −6.53, −6.61, −7.21 (Si(CH_3)); MS (70 eV): m/z (%) = 486 (4.5, M^+).

Ferrocenylundecamethylcyclohexasilane (**4**)

The procedure followed was that used for 1-ferrocenyl-4-(4-bromophenyl)-decamethylcyclohexasilane with 5.0 g (13.5 mmol) of chloroundecamethylcyclohexasilane, 5.8 g (31.4 mmol) of ferrocene, and 9 mmol of *tert*-butyllithium in 10 cm^3 of *THF*.

Yield: 5.6 g (80% based on chloroundecamethylcyclohexasilane); m.p.: 118–119°C; $C_{21}H_{42}FeSi_6$ (518.93); found: C 48.48, H 8.31; C 48.60, H 8.15; ^{29}Si NMR (Toluol/D_2O, ext. *TMS*, 59.62 MHz): $\delta = -41.75$, −41.78, −41.83, −42.20 ppm; ^{1}H NMR (C_6D_6, ext. *TMS*, 300.13 MHz): $\delta = 4.18$ (2H, ferrocene), 4.09 (s, 5H, ferrocene), 4.00 (2H, ferrocene), 0.43 (s, 3H, Si(CH_3)); 0.26 (s, 3H, Si(CH_3)); 0.24 (s, 3H, Si(CH_3)); 0.22 (s, 3H, Si(CH_3)); 0.21 (s, 3H, Si(CH_3)); 0.15 (s, 6H, Si(CH_3)$_2$) ppm; ^{13}C NMR (C_6D_6, ext. *TMS*, 75.47 MHz): $\delta = 73.56$, 70.63, 68.40 (ferrocene), −4.44, −5.54, −5.63, −6.38, (Si(CH_3)) ppm; MS (70 eV): m/z (%) = 518 (49, M^+).

Acknowledgements

We thank the *Fonds zur Förderung der wissenschaftlichen Forschung* (Wien, Austria) for financial support within the *Spezialforschungsbereich Elektroaktive Stoffe* and the *Wacker Chemie GmbH* (Burghausen, Germany) for the donation of silane precursors.

References

[1] Wenjiang N (1993) Adv Mater **5**: 520

[2] a) Williams DJ (1984) Angew Chem **96**: 637; b) Long NJ (1995) Angew Chem **107**: 37; c) Whittall IR, McDonagh AM, Humphrey MG, Samoc M (1999) Adv Organomet Chem **42**: 291; d) Whittall IR, McDonagh AM, Humphrey MG, Samoc M (1999) Adv Organomet Chem **43**: 349; e) Marder SR, Sohn JE, Stucky GD (eds) (1991) Materials for Nonlinear Optics: Chemical Perspectives. ACS Symposium Series 455, American Chemical Society, Washington, DC; f) Chemla DS, Zyss J (eds) (1987) Nonlinear Optical Properties of Organic Molecules and Crystals. Academic Press, Orlando, FL

[3] Marder SR, Perry JW, Tiemann BG (1991) Organometallics **10**: 1896

[4] Green MLH, Marder SR, Thompson ME, Bandy JA, Bloor D, Kolinsky PV (1987) Nature **330**: 360

[5] a) Mignani G, Barzoukas M, Zyss J, Soula G, Balegroune F, Grandjean D, Josse D (1991) Organometallics **10**: 3660; b) Mignani G, Krämer A, Puccetti G, Ledoux I, Soula G, Zyss J, Meyrueix R (1990) Organometallics **9**: 2640; c) Mignani G, Krämer A, Puccetti G, Ledoux I, Zyss J, Soula G (1991) Organometallics **10**: 3656; d) Hissink D, van Hutten PF, Hadziioannou G (1993) J Organomet Chem **454**: 25; e) Hissink D, Brouwer J, Flipse R, Hadziioannou G (1991) Polymer Pepr. **32**: 136; f) Hissink D, Bolink HJ, Eshuis JW, Malliaras GG, Hadziioannou G (1993) Polymer Pepr **34**: 721

[6] a) Sakurai H (1980) J Organomet Chem **200**: 261; b) West R (1986) J Organomet Chem **300**: 327; c) West R (1995) In: Wilkinson G, Stone FGA, Abel EW (eds) Comp Organomet Chem II, vol 2. Pergamon Press, Oxford, p 77; d) Patai S, Rappoport Z (eds) (1989) The Chemistry of Organic Silicon Compounds. Wiley, Chichester; e) Patai S, Rappoport Z (eds) (1998) The Chemistry of Organic Silicon Compounds, vol 2. Wiley, Chichester; f) Pitt CG (1977) In: Rheingold AL (ed) Homoatomic Rings, Chains and Macromolecules of Main Group Elements. Elsevier, New York, p 203

[7] Sharma HK, Pannell K, Ledoux I, Zyss J, Ceccanti A, Zanello P (2000) Organometallics **19**: 770

[8] Grogger C, Siegl H, Rautz H, Stüger H (2000) In: Auner N, Weis J (eds) Organosilicon Chemistry 4. Verlag Chemie, Weinheim p 384

[9] West R (1982) Pure Appl Chem **54**: 1041

[10] Thomas KRJ, Lin JT, Wen YS (1999) J Organomet Chem **575**: 301

[11] Compare Ref [7], p 774

[12] Helmer BJ, West R (1982) J Organomet Chem **236**: 21

[13] Compare the explanation for the electron releasing character of the $(CH_3)_3Si$ group in Ref. 5a including bibliography

[14] EFISH measurements are performed in the laboratories of J. Zyss; detailed results will be published in a subsequent paper

[15] Rebiere F, Samuel O, Kagan HB (1990) Tetrahedron Lett **31**: 3121

[16] Hengge E, Eibl M, Schrank F (1991) Spectrochim Acta **47a**: 721

[17] Spielberger A, Gspaltl P, Siegl H, Hengge E, Gruber K (1995) J Organomet Chem **499**: 241

[18] Hengge E, Eibl M (1992) J Organomet Chem **428**: 335

[19] Gagné RR, Koval CA, Lisensky GC (1980) Inorg Chem **19**: 2855

[20] Shafiq FA, Richardson DE, Boncella JM (1998) J Organomet Chem **555**: 1

Received May 30, 2000. Accepted December 20, 2000

Thin-Film Zinc/Manganese Dioxide Electrodes

Paul Barbic[1], **Leo Binder**[1,*], **Susanne Voß**[1], **Ferdinand Hofer**[2], and **Werner Grogger**[2]

[1] Institute of Chemical Technology of Inorganic Materials, Technical University Graz, A-8010 Graz, Austria
[2] Research Institute of Electron Microscopy, A-8010 Graz, Austria

Summary. Thin-film electrodes allow the manufacturing of flat batteries of variable design. Their electric performance is better than that of customary cells because of a larger contact area between anode and cathode and better utilization of the electrochemically active materials. Bipolar thin-film electrodes for the rechargeable alkaline zinc/manganese dioxide system were assembled using graphite-filled plastics (high-density polyethylene and polyisobutylene). In a different approach, extremely thin electrodes were obtained using 25 µm thick foils made of non-conductive micro-porous polypropylene. The electroactive materials were electrolytically deposited into the pores of the previously metallized foil, providing the required conductive connection through the plastic matrix by themselves. Cycle behavior, cumulated capacities, and energies of batteries with up to two bipolar units were measured. At this early stage of development, batteries based on graphite-filled polymer foils showed better results with regard to storage capacity per unit area and to cycle life. Prototypes based on micro-porous polypropylene suffered from the relatively small fraction of pore volume available for the deposition of active material (about 38%) and from current collector corrosion.

Keywords. Rechargeable alkaline zinc/manganese dioxide batteries; Thin-film electrodes; Bipolar electrodes.

Introduction

An important advantage of thin-film electrodes is a better performance than that of customary (standard) cells because of a larger contact area between anode and cathode, a better utilization of the electrochemically active materials [1], and shorter diffusion paths. Substituting the metal carrier by conductive plastics lead to a considerable reduction of battery weight (factor 10 using the Celgard® foil mentioned below compared with sheet brass). The aggravating disadvantages of using plastics as carrier for the electroactive materials are their low conductivity and their lack of adherence to the electroactive layers. Conductivity may be improved by incorporation of conductive fillers like graphite or/and soot in the polymer.

Usually, the electroactive materials are deposited onto the surface of a conductive carrier (metal or conductive polymer, respectively). A significant improve-

* Corresponding author

ment of the originally poor adherence between polymer foil and electroactive layers can be achieved by mechanical (surface roughening) or chemical (etching) pretreatment [2–4]. Using plastics with conductive fillers, like graphite and/or soot incorporated in the polymer, demands for an intermediate layer between the carbon-filled polymer and zinc to prevent formation of hydrogen gas by anodic zinc corrosion, which is enhanced in the presence of carbon in any modification.

Attempts to use intrinsically conductive polymers like polypyrrole as carrier materials require special precautions and manufacturing processes because most of these polymers are insoluble and brittle. Thus, further treatment is hardly practicable. The advantage of intrinsically conductive polymers compared with filled plastics is that an intermediate layer between carrier and zinc could be avoided.

The achievable resistance of these materials (plastics with conductive fillers and intrinsically conductive polymers) is of course not comparable to that of metal carriers, but it is sufficient for thin foils in bipolar arrangements.

Results and Discussion

Graphite filled carriers

The cycle behavior of battery stacks with up to two bipolar units was investigated. Figures 1–4 compare the results of batteries based on high-density polyethylene with those based on polyisobutylene. The 4.65 V batteries were discharged at 50 mA (constant current) until a cut-off voltage of 900 mV was reached. Thereafter, they were charged at 5400 mV (constant voltage) for eight hours.

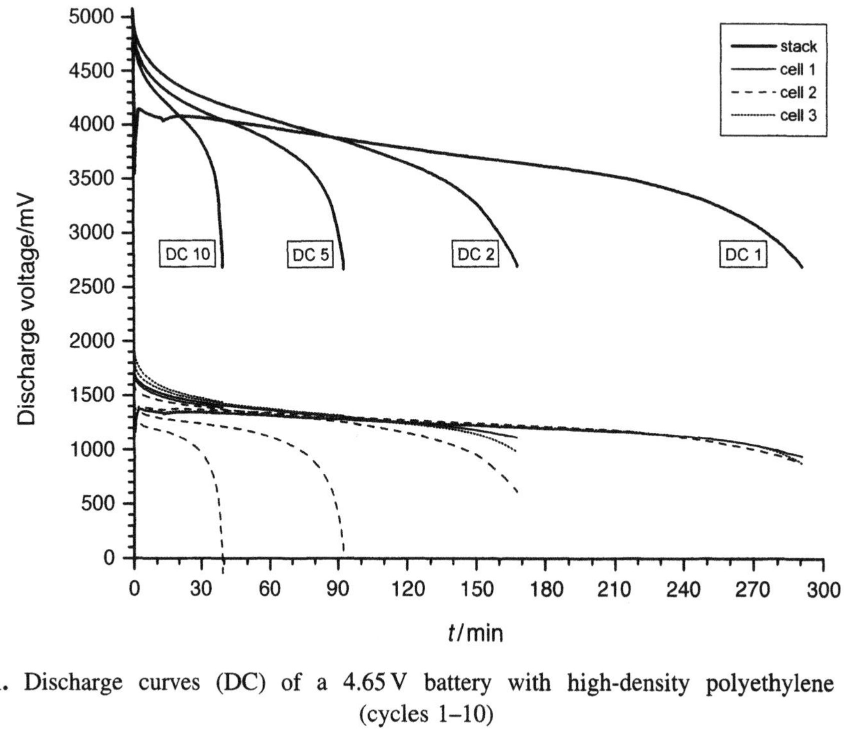

Fig. 1. Discharge curves (DC) of a 4.65 V battery with high-density polyethylene carriers (cycles 1–10)

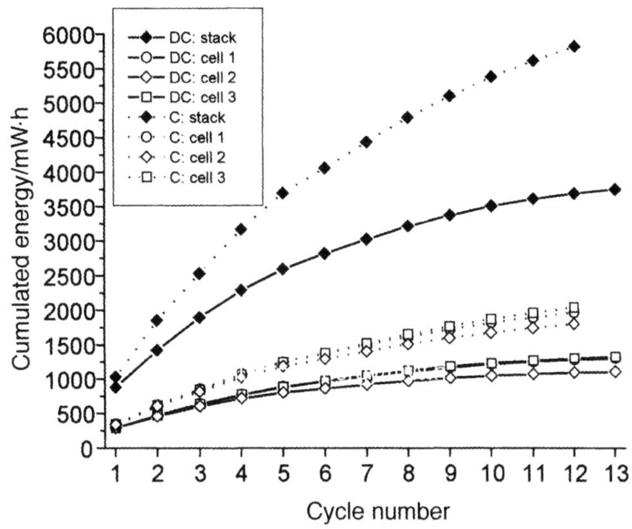

Fig. 2. Cumulated discharge (DC) and charge (C) energies of a 4.65 V battery stack containing 2 bipolar units + terminal cathode + terminal anode (= 3 cells) with high-density polyethylene carriers

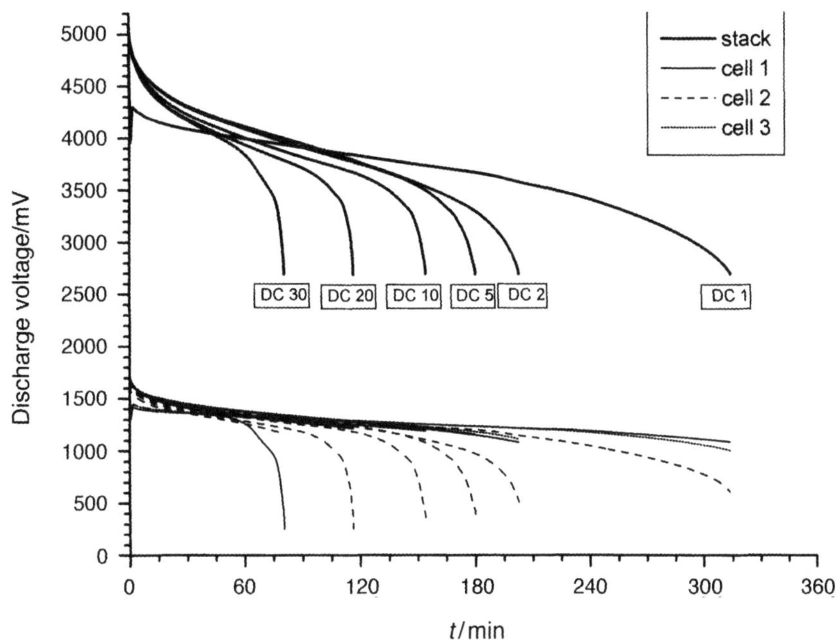

Fig. 3. Discharge curves of a 4.65 V battery with polyisobutylene carriers (cycles 1–30)

Porous polypropylene as carrier

To check the degree of filling of the pores, scanning electron microscope (SEM) photographs and analyses were obtained from a cryosection at the research institute of electron microscopy. They proved an entire filling of the pores and a 5–10 μm thick surplus layer on the surface. X-Ray diffraction was used to confirm that γ-MnO_2 was deposited [4].

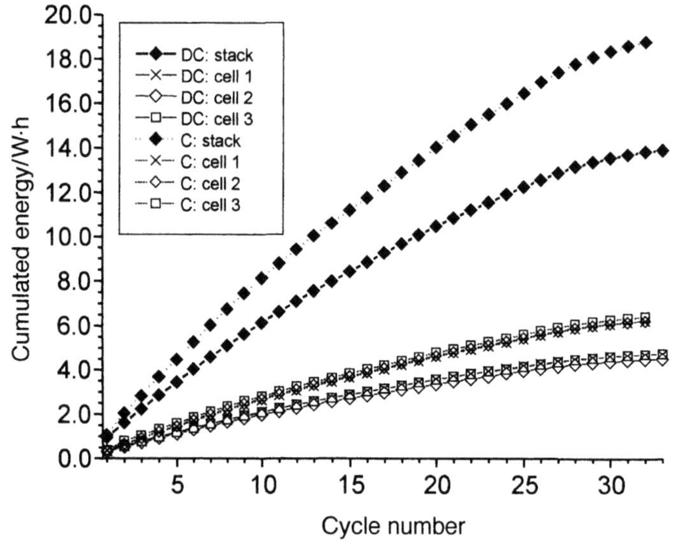

Fig. 4. Cumulated discharge (DC) and charge (C) energies of a 4.65 V battery stack containing 2 bipolar units + terminal cathode + terminal anode (= 3 cells) with polyisobutylene carriers

The separator between cathode and anode must provide enough capacity to store the electrolyte (KOH) and prevent the growth of zinc dendrites from the zinc anode to the cathode during charging. Thus, one side of such a separator should be made of a non-porous material like cellophane.

Studies on the DC-behavior of Celgard®-based zinc/manganese dioxide cells showed a rather high corrosion rate of the copper current collector and KOH leakage. The potassium hydroxide takes part in the anodic electrode reaction. Therefore, KOH leakage affects adversely cycle life as well as shelf life of the battery and had to be minimized. Unfortunately, the porous structure of the polymeric carrier enhances the property of KOH to creep into every crevice. Thus, the edges of the electrodes were pasted very well using an epoxy resin. Cells without and with bipolar electrodes (cell-stacks) were investigated with respect to their behavior during charging and discharging. Discharge was stopped at a COV (cut-off voltage) of 900 mV (without bipolar unit) and 1800 mV (with one bipolar unit), respectively. After discharging had finished, the cell was charged potentiostatically with 1700 mV (without bipolar unit) and 3400 mV (with one bipolar unit) for three hours. Thereafter, discharge was performed again.

To achieve a more corrosion resistant current collector for the anodic side (copper will oxidize when zinc is completely dissolved during discharge), a thin tin layer was deposited before zinc deposition. Analysis showed that there was no individual tin layer, but tin diffused into the copper layer. Using this copper-tin alloy as current collector, the cycle life of assembled cells not only increased from average four to average 20 cycles, but, at least for the first four cycles, the cumulated capacity on discharge almost doubled from 2.4 mA h to 3.9 mA h. (Figs. 5 and 6). Due to the oxidation of the less corrosion resistant copper, a current collector layer made of this material disintegrates easily. Therefore, parts of the

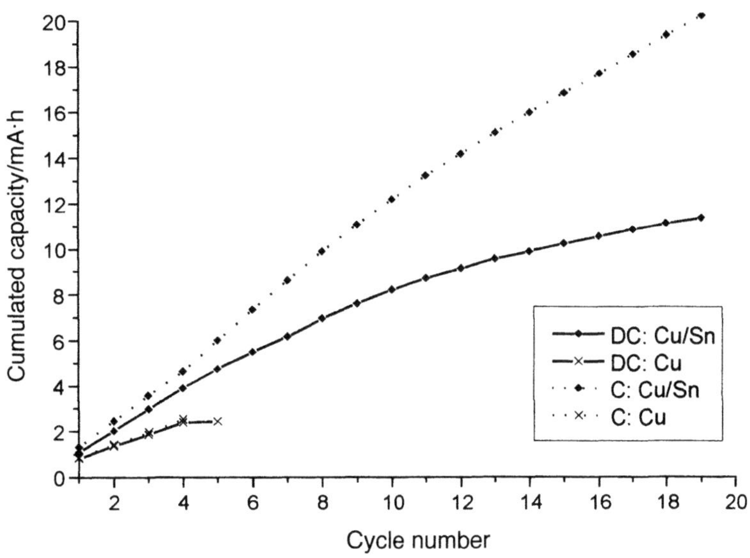

Fig. 5. Influence of the current collector material on the cumulated capacity of a 1.5 V cell based on porous polypropylene carrier (3 mA discharge current, 9 cm^2 electrode surface)

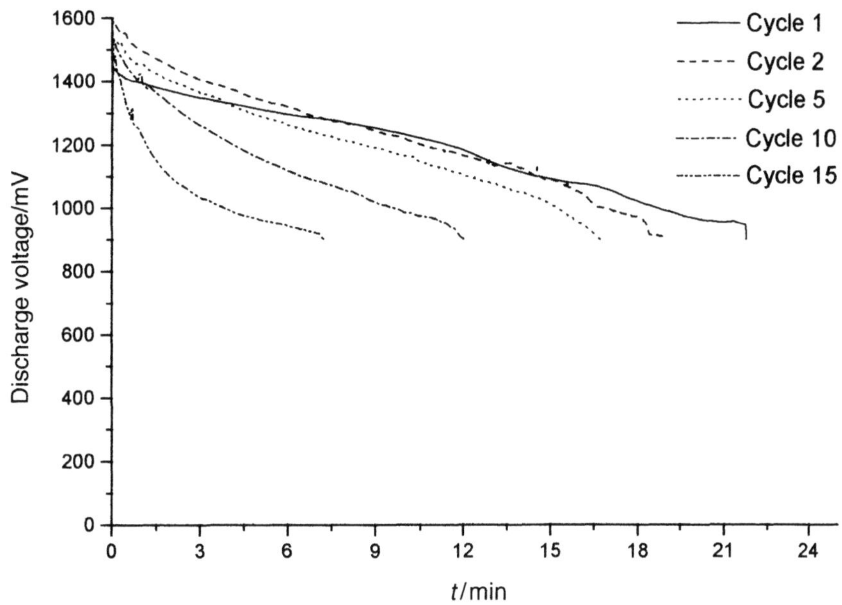

Fig. 6. Discharge curves for a zinc/manganese dioxide cell with Celgard® foils as carriers (discharge current: 3 mA, electrode surface: 9 cm^2)

negative electrode cannot be recharged again, causing less capacity in the next discharge cycle.

An external reinforcement of these cells had a significant influence on the OCV and CCV (closed circuit voltage), respectively, because the polymer itself is not as stable as metallic carriers. Consequently, contact problems will appear, particularly if the internal pressure increases due to *e.g.* zinc corrosion. Thus, a stable casing is needed to press the cell-stack together.

First attempts were made to achieve thin-film bipolar electrodes by assembling these electrodes in such a way that their metallized sides stick together.

Conclusions

Bipolar batteries using graphite filled carriers showed the well known phenomena of layer separation (active masses partly lost contact to the carrier) during repeated cycling. Generally speaking, due to the simple production process, the easy handling of bipolar units, and the higher amount of active mass per unit area, the performance of batteries with graphite filled carriers was superior to that of batteries using porous polypropylene carriers.

On the other hand, in the case of porous polypropylene carriers all effects of contact loss and layer separation were completely suppressed by the perfect inclusion of the active masses. The amount of active mass was limited by the open pore volume available for electrochemical deposition (only 38%). This fact explains the low storage capacity per unit area in comparison to prototypes using conventional graphite-filled polymer foils (ratio about 1:4). An intensive search for foils with equal thermal and chemical stability but significantly higher porosity seems to be inevitable.

All described electrodes were made from standard materials (zinc and manganese dioxide), and performance may be improved by changing to particular materials designed for rechargeable batteries (alloyed zinc, corrosion inhibiting additives, modified manganese dioxides, additives increasing conductivity).

Experimental

Each of the above mentioned polymers must comply with the demands of electrolytic deposition of the electroactive materials (*e.g.* 90°C, 1 mol/dm^3 H_2SO_4 bath, anodic polarization for deposition of manganese dioxide) as well as with the requirements of battery operation using 9 mol/dm^3 potassium hydroxide as cell electrolyte.

Electrodes based on carbon-filled plastics (high-density polyethylene (*HDPE*) and polyisobutylene (*PIB*)) and electrodes based on micro-porous polypropylene foil (Celgard®), respectively, were investigated with respect to cycle behavior, cumulated capacity, and energy output.

Graphite filled carriers

HDPE and *PIB*, respectively, were pressed to 0.2 mm thin layers using a Carver laboratory press at temperatures from 170 to 190°C. A 9 μm thin tin layer (previously etched in conc. HCl for ten seconds) was pressed onto the anodic side of the softened plastic to avoid zinc corrosion due to contact with carbon. After roughening of the cathodic side by pressing a grinding paper onto the softened plastics for about 15 seconds, a thin graphite layer was sprayed onto the roughened side of the electrode, which increases the electric conductivity at the contact area between plastics and manganese dioxide. A zinc layer was applied onto the tin-plated side by silk-screen printing. In addition, a separator was bonded to this side employing an epoxy-based adhesive. Thereafter, a mixture of 90% MnO_2, 9.5% graphite, 0.5% soot, and binder was applied onto the roughened electrode's surface by silk-screen printing [4].

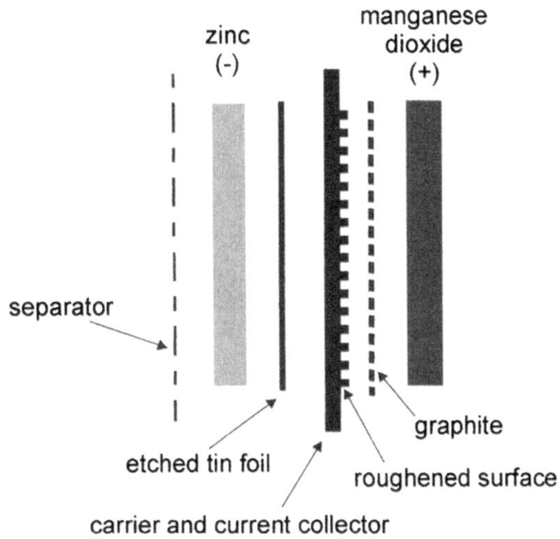

Fig. 7. Schematic assembly of a bipolar electrode based on graphite-filled carriers

The advantage of the *PIB*-carbon composite material was mainly the availability of thin foils (0.2 mm and lower). *HDPE*-carbon sheets had an original thickness of about 1 mm, and the pressing process was of higher importance. The advantage of *HDPE* in comparison to *PIB* was the slightly lower softening point. Surface properties and conductivity of both materials were similar.

Cells with a theoretical open-circuit voltage calculated from Eq. (1) were achieved by assembling these bipolar electrodes to stacks. 9 *M* KOH solution was used as electrolyte. Figure 7 shows a schematic assembly.

$$OCV_{\text{theor}} = 1.55\,\text{V} \cdot (\text{number of bipolar electrodes} + 1) \qquad (1)$$

Porous polypropylene as carrier

The metallization of the polypropylene foil (*PP*-foil) Celgard®-2402 followed a modified commercial coating process [5, 6]. The metals are deposited within the microporous structure and on the surface of the *PP*-foil. A well-adherent metal layer at the substrate Celgard®-2402 results which is mechanically anchored in the pores by metallic roots. The penetration depth of the metal layer can be controlled by the added surfactants in the electroplating baths.

Two steps are necessary for the electroless metallization. In the first step, the surface of the *PP*-foil will be catalytically activated by finely dispersed palladium crystal nuclei. After activation, the substrate will be electrolessly coated with the corresponding metal (copper or palladium).

Whereas copper was used as negative current collector, only palladium seemed to be usable as positive counterpart because of the conditions during electrolytic deposition of manganese dioxide (temperature: 88–93°C, H_2SO_4 bath, anodic polarization). The described electrode prototypes were equipped with a palladium layer of roughly 1 μm (equivalent to about $1\,\text{mg} \cdot \text{cm}^{-2}$ Pd). In future experiments this amount will have to be decreased for cost reasons.

After electroless metallization, the two originally attached layers of the Celgard®-2402 foil were pulled apart to achieve two single-side metallized polymer layers. These can be used as thin-film electrodes in rechargeable alkaline zinc/manganese dioxide (RAM™) batteries. The electroactive materials, zinc and manganese dioxide, were electrochemically deposited in the pores of the single-side metallized and moistened *PP*-foil by means of pulse electrolysis. A schematic production process is shown in Fig. 8, and a detailed description of electrolysis parameters has been given in Refs. [4, 7].

After washing with distilled water the electrodes were assembled to batteries. Berec (contains a layer of cellophane) was used as separator, and 9 *M* KOH was employed as cell electrolyte.

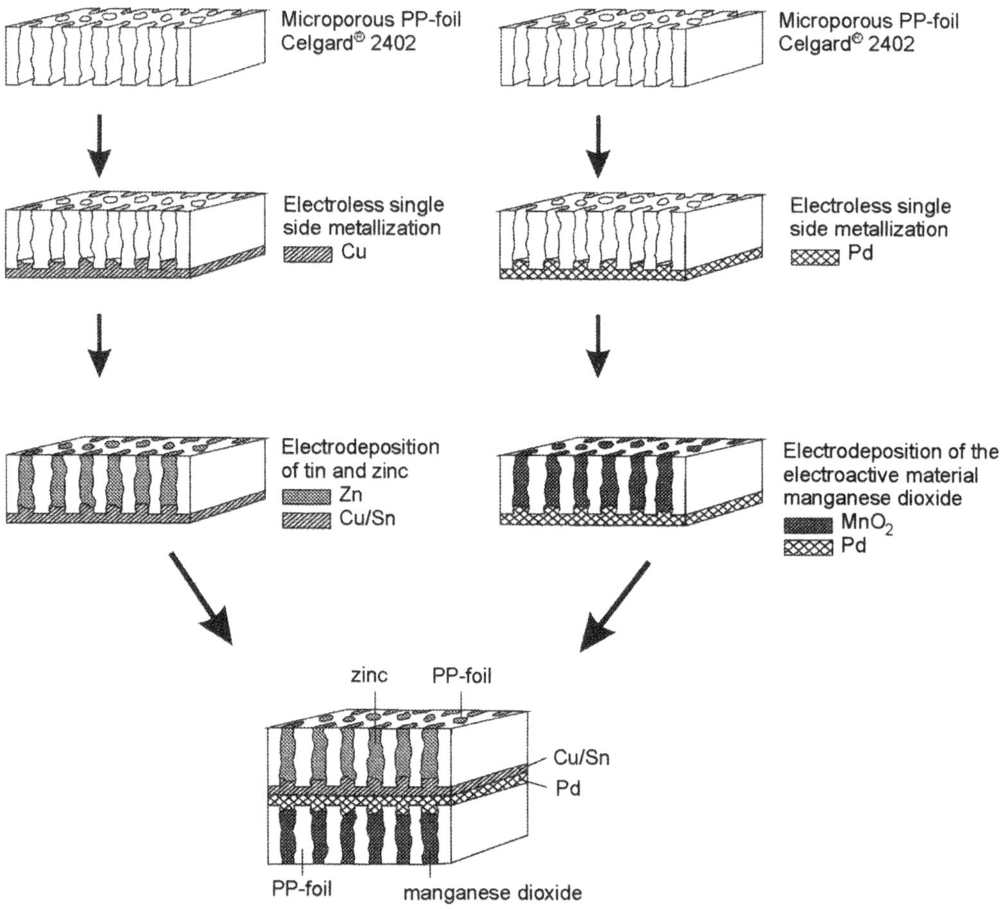

Fig. 8. Production process of bipolar thin-film zinc/manganese dioxide electrodes based on microporous polyethylene carriers

Acknowledgements

Grateful acknowledgement is made to the Special Research Area *Electroactive Materials*, Austrian Research Foundation, for financial support of this work.

References

[1] Binder L, Kordesch K (1993) DECHEMA Monographien **128**: 313
[2] Ghaemi M (1995) Dissertation, Technical University Graz, Austria
[3] Barbic PA (1997) Master Thesis, Technical University Graz, Austria
[4] Barbic PA (1999) Dissertation, Technical University Graz, Austria
[5] Besenhard JO, Michels L, Huslage J (1996) DT PA 42 14 905
[6] Besenhard JO, Heß M, Huslage J, Krebber U (1993) Power Sources **43–44**: 493
[7] Barbic PA, Binder L, Voss S, Hofer F, Grogger W (1999) J Power Sources **79**: 271

Received May 30, 2000. Accepted December 18, 2000

Studies on the Anode/Electrolyte Interface in Lithium Ion Batteries

Martin Winter[1,*], **Wolfgang K. Appel**[2], **Bernd Evers**[1], **Tomáš Hodal**[1], **Kai-Christian Möller**[1], **Ingo Schneider**[1], **Mario Wachtler**[1], **Markus R. Wagner**[1], **Gerhard H. Wrodnigg**[1], and **Jürgen O. Besenhard**[1]

[1] Institute for Chemical Technology of Inorganic Materials, Graz University of Technology, A-8010 Graz, Austria

[2] Aventis Research and Technologies GmbH & Co KG (former Hoechst Research and Technology Deutschland), Operative Research, D-65926 Frankfurt am Main, Germany

Summary. Rechargeable lithium ion cells operate at voltages of 3.5–4.5 V, which is far beyond the thermodynamic stability window of the battery electrolyte. Strong electrolyte reduction and anode corrosion has to be anticipated, leading to irreversible loss of electroactive material and electrolyte and thus strongly deteriorating cell performance. To minimize these reactions, anode and electrolyte components have to be combined that induce the electrolyte reduction products to form an effectively protecting film at the anode/electrolyte interface, which hinders further electrolyte decomposition reactions, but acts as membrane for the lithium cations, *i.e.* behaving as a solid electrolyte interphase (SEI). This paper focuses on important aspects of the SEI. By using key examples, the effects of film forming electrolyte additives and the change of the active anode material from carbons to lithium storage alloys are highlighted.

Keywords. Anode; Graphite; Lithium alloy; Lithium ion battery; Solid electrolyte interphase.

Introduction

Numerous rechargeable lithium cells have been developed in the last decades in order to satisfy the increasing demands for high energy density batteries for portable electronic applications. Only few of them have been successfully commercialized, the most important to date being the lithium ion battery. The high specific energies (>130 $Wh \cdot kg^{-1}$) and energy densities (>250 $Wh \cdot dm^{-3}$) of lithium ion cells make their application highly attractive. Beginning with the introduction of lithium ion cells in 1990 [1], their share in the rechargeable consumer battery market has dramatically increased over the years. The estimated world wide production was 200 million cells for 1997 [2]. Lithium ion cells play a major role in the so called "3C" market (cellular phones, portable computers, camcorders). Moreover, several

* Corresponding author

companies have announced or established a prototype production of electric vehicles equipped with large lithium ion batteries. More than 90% of the cells are manufactured in Japan. There are also strong activities in Europe and in the US to participate in this market, either by manufacture of cells or by production of battery materials. Moreover, in late 1999/early 2000 several manufacturers will be instituted in China, Korea, and Taiwan, which produce lithium ion cells at a much lower cost [3].

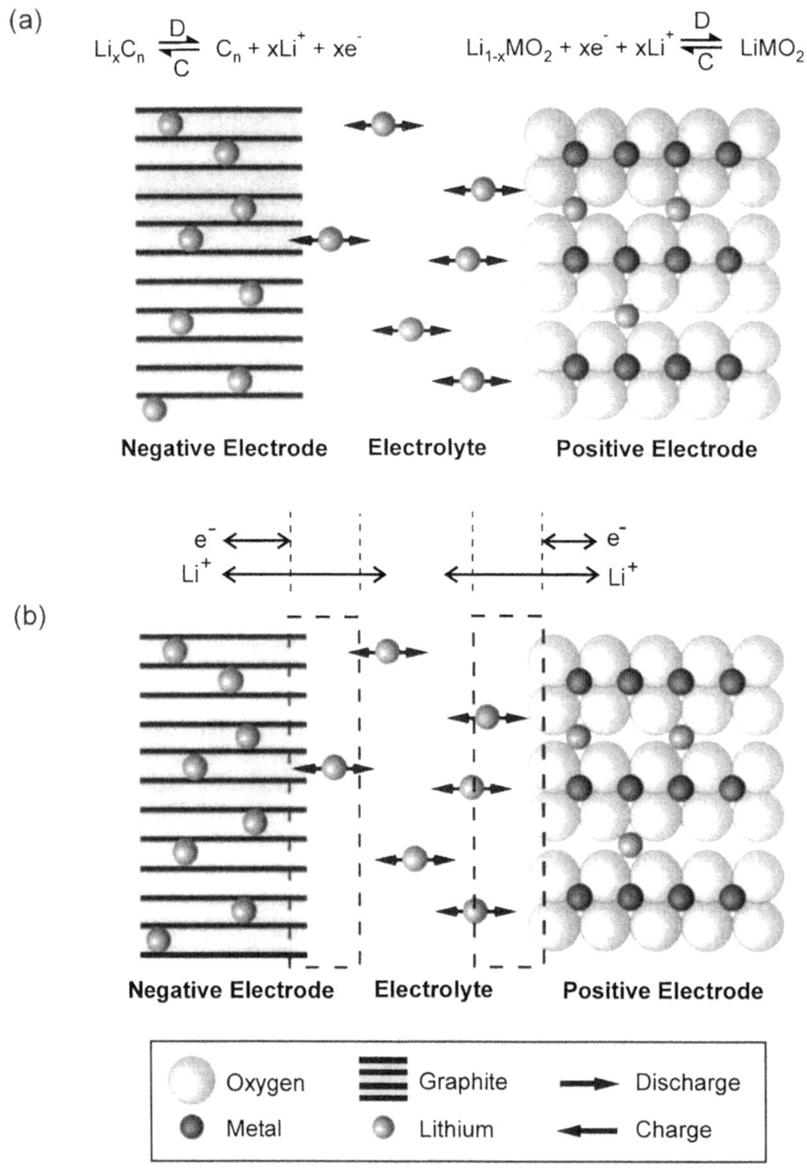

Fig. 1. (a) Schematic representation of a lithium ion cell and the charge/discharge reactions at the electrodes; as for the graphite negative electrode (anode), layered insertion electrode materials, *e.g.* $LiCoO_2$ and $LiNiO_2$, are used for the positive electrode (cathode) as well; (b) formation of electronically insulating but lithium ion conducting interphases at the electrodes

In lithium ion cells no metallic lithium is present at any state of charge/discharge. Both electrodes are capable of reversible lithium insertion (Fig. 1a). Basically, this insertion reaction is a host/guest solid state redox reaction involving electrochemical charge transfer coupled with insertion of mobile guest ions from an electrolyte into the structure of a solid host, which is a mixed electronic and ionic conductor. In commercial cells, mainly carbonaceous hosts (non-graphitic carbons or graphites) are used for the negative electrode, and lithium/transition metal oxides are employed for the positive electrode ($LiCoO_2$, $LiNiO_2$, or $LiMn_2O_4$). Because of the strong difference of the chemical potential of lithium in the two electrodes, which corresponds to cell voltages of more than 3.5 V, the transfer of lithium ions from the negative electrode through the electrolyte to the positive electrode (discharge) delivers energy whereas the reverse lithium transfer (charge) consumes energy. In view of electrochemical performance, as well as economical and ecological impacts, $LiMn_2O_4$ cathodes and graphitic anodes seem to be the preferred electrode materials for lithium ion batteries in the future. Typically, lithium salts ($LiClO_4$, $LiBF_4$, $LiPF_6$, $LiN(SO_2CF_3)_2$, *etc.*) dissolved in aprotic organic solvents, *e.g.* esters or ethers or mixtures thereof, are used as electrolytes [4, 5].

The performance of lithium ion batteries strongly depends on the type of electrode material [4]. To a first approximation, properties such as charge storage capacity (denoted as specific charge ($Ah \cdot kg^{-1}$) or charge density ($Ah \cdot dm^{-3}$)), redox potentials for lithium insertion/de-insertion, reversibility, and cycle life determine the choice of a certain material. Since the introduction of lithium ion cells to the market, the specific energy has increased by more than 30%, mostly due to the use of carbon anode materials with higher lithium storage capabilities. Many manufacturers use graphites as anode materials, which typically exhibit a maximum lithium capacity of LiC_6 (339 $Ah \cdot kg^{-1}$/759 $Ah \cdot dm^{-3}$ with respect to lithiated graphite and 372 $Ah \cdot kg^{-1}$/837 $Ah \cdot dm^{-3}$ with respect to unlithiated graphite). However, there is strong interest to replace carbons by anode materials which can show even higher specific charges/charge densities. A large number of metals and intermetallics ("alloys"), such as Al, Si, Sn, Sb, "SnSb", $SnAg_3$ *etc.*, are capable of reversible accommodation of lithium. These lithium storage metals and alloys show very high specific charges and charge densities (*e.g.* $Li_{22}Sn_5$: 790 $Ah \cdot kg^{-1}$/2020 $Ah \cdot dm^{-3}$) and have therefore been repeatedly suggested as anode materials for Li ion batteries [6]. Unfortunately, the uptake and release of Li is accompanied by enormous volume changes (*e.g.* from Sn to $Li_{22}Sn_5$: approx. 250% volume increase; by comparison, from graphite to LiC_6: only approx. 10% volume increase), which in the case of ordinary coarse-grained, bulky metal host materials leads to cracking and crumbling of the electrode and hence renders an application in rechargeable batteries impossible [6] (see Results and Discussion).

Lithium ion cells exhibit cell voltages of up to 4.5 V and therefore operate far beyond the thermodynamic stability window of the organic electrolytes. Electrolyte decomposition occurs at both electrodes. Fortunately, electrolyte reduction products, created *in situ* during charge, form protecting films at the negative electrode which – in the ideal case – are electronically insulating and thus hinder further electrolyte reduction but still act as a membrane for the active charge carrier, the lithium cation (Fig. 1b). In other words, these films behave as a solid electrolyte interphase (SEI) [7][1]. Since the SEI formation is associated with the irreversible consumption of material (lithium and electrolyte), the corresponding charge loss is called irreversible capacity, C_{irr}[2]. The irreversible capacities have to be minimized because they are detrimental to both specific energy and energy density of the cell and, moreover, increase the material expenses due to the necessary excess of costly positive electrode material which is the lithium source after cell assembly. In the following we will show some basic strategies which are pursued within the special research program "Electroactive Materials" in order to reach this goal. By way of illustrative examples, special emphais will be put on new electrolyte additives and reactions taking place at the lithium storage alloy interface.

Results and Discussion

Novel electrolyte solvents and electrolyte additives

Mixed solvent electrolytes containing highly viscous ethylene carbonate (*EC*) and low viscosity dilutants such as dimethyl carbonate (*DMC*) or diethyl carbonate (*DEC*) as main solvent components are presently used in commercial lithium ion batteries comprising graphitic carbons for the anode. The structural formulae of these and other electrolyte solvents and additives reported in this study are shown in Fig. 2. *DMC* and/or *DEC* are required to get a reasonable low temperature performance of the electrolyte. *EC* is indispensable because of its excellent anode filming properties. In particular, solvent co-intercalation into graphitic anodes, which is observed in many electrolytes, can be suppressed in the presence of sufficient amounts of *EC*. The massive co-intercalation of the large solvent molecules into the interlayer gaps of the graphite matrix usually leads to a drastic volume increase (>100%) which often results in electrode destruction. Moreover, the co-intercalated solvent molecules are in contact with surfaces inside the graphite. These surfaces then also take part in the film formation process. This reaction considerably increases the undesired C_{irr} in comparison to SEI formation limited to the external graphite surfaces [4, 5].

Electrolyte additives

The use of *EC*-based electrolytes involves certain limitations in bulk electrolyte properties such as poor low-temperature conductivity and high flammability. Many research groups have therefore proposed new electrolyte components which still ensure the formation of the desired SEI, but show improved bulk electrolyte

[1] The SEI concept describes the filming behaviour observed on the Li_xC_6 negative electrode in a very basic and general way. Though the function and the formation process of the SEI in several cases seems to be more complex, the concept has been quite generally accepted due to its simplicity. The situation is different at the positive electrode. Up to now, there are only a few reports on the formation and composition of protective interphases at the positive electrode (Fig. 1b and, *e.g.* Ref. [4]), and the nature and *modus operandi* of these films is still not clear.

[2] The reversible lithium insertion, on the other hand, is called discharge capacity or reversible capacity, C_{rev}.

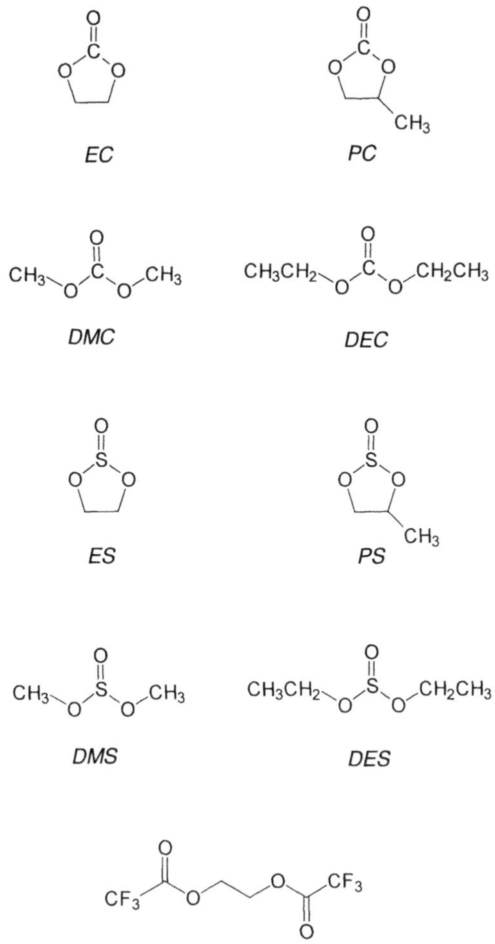

Fig. 2. Structural formulae of ethylene carbonate (*EC*), propylene carbonate (*PC*), dimethyl carbonate (*DMC*), diethyl carbonate (*DEC*), ethylene sulfite (*ES*), propylene sulfite (*PS*), dimethyl sulfite (*DMS*), diethyl sulfite (*DES*), and 1,2-*bis*-(trifluoracetoxy) ethane (*BTE*)

properties. In order to decouple the filming and bulk properties of the electrolyte, reactive electrolyte additives can be used [8, 9]. Even in small amounts these additives ensure a quick formation of the SEI, thus allowing the selection of the main electrolyte component independently of its filming properties.

The *modus operandi* of such electrolyte additives will be illustrated by ethylene sulfite (*ES*) used as an additive in a propylene carbonate (*PC*) electrolyte. Electrolytes based on *PC* are known for their better low-temperature behaviour compared to *EC* based ones. Unfortunately, strong *PC* co-intercalation into graphite takes place, which is followed by reduction of the solvated intercalates, $Li(PC)_yC_n$. One main reduction product is propene gas [10]. Due to *PC* co-intercalation and gas formation inside the graphite, shedding of single graphene layers or packages of graphene layers (exfoliation) occurs. Eventually, the electrode is destroyed. This process, simply called solvated intercalation, usually starts at electrode potentials of

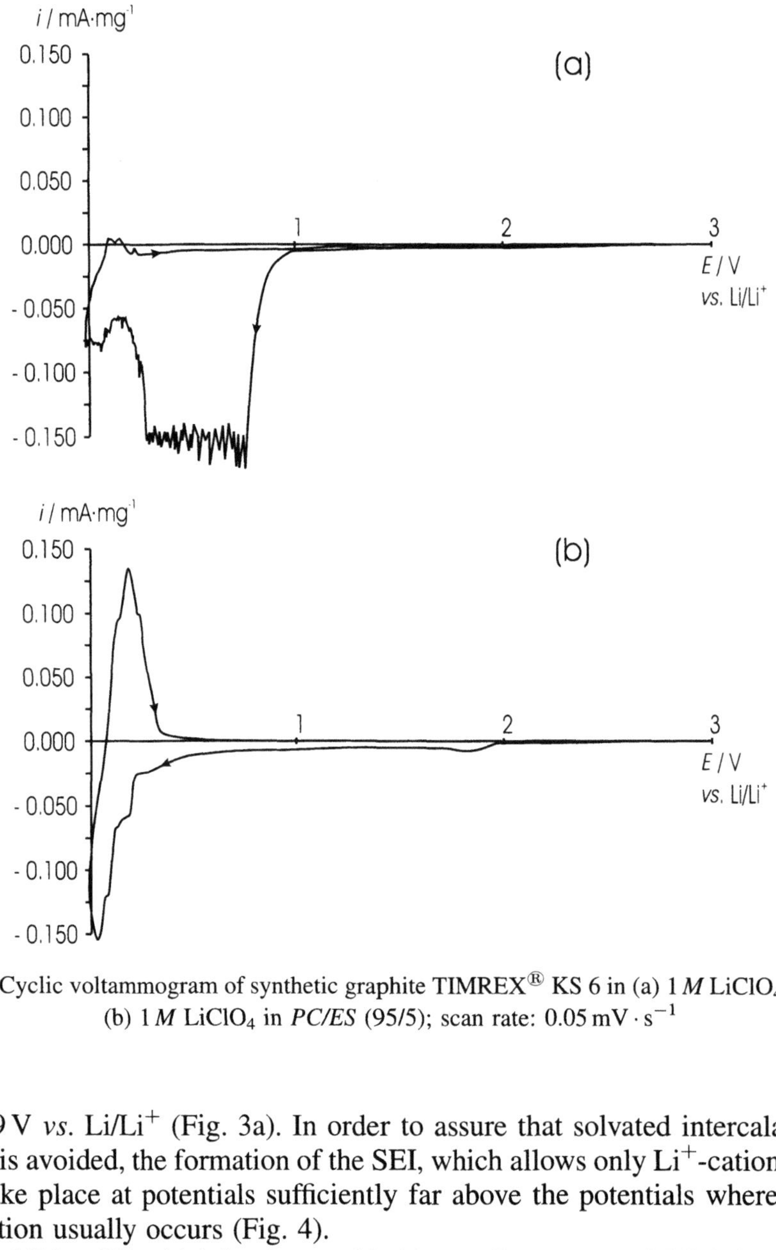

Fig. 3. (a) Cyclic voltammogram of synthetic graphite TIMREX® KS 6 in (a) 1 *M* $LiClO_4$ in *PC* and (b) 1 *M* $LiClO_4$ in *PC/ES* (95/5); scan rate: 0.05 mV · s^{-1}

~1.0–0.9 V *vs.* Li/Li^+ (Fig. 3a). In order to assure that solvated intercalation into graphite is avoided, the formation of the SEI, which allows only Li^+-cations to pass, has to take place at potentials sufficiently far above the potentials where solvated intercalation usually occurs (Fig. 4).

The additive *ES*, which has been added in small amounts of 5% (v/v) to the *PC* electrolyte, is reduced at electrode potentials of ~2.0 V *vs.* Li/L^+. The reduction products obviously form a SEI, which inhibits *PC* co-intercalation. This is confirmed by the absence of current peaks at ~1 V *vs.* Li/Li^+ in the voltammogram (Fig. 3b). No exfoliation of graphite occurs. Instead, the desired reactions, intercalation and de-intercalation of unsolvated lithium, are observed at potentials of ~0.3–0.0 V *vs.* Li/Li^+. Organic sulfites, whether cyclic (ethylene sulfite [8], propylene sulfite (*PS*) [11]) or linear (dimethyl sulfite (*DMS*), diethyl sulfite (*DES*) [12]) basically show a very similar filming behavior. However, the filming strength on graphitic anodes is different, *i.e.* the formation of the SEI is associated with different extents of C_{irr}. It can be estimated as follows: $ES > PS \gg DMS > DES$,

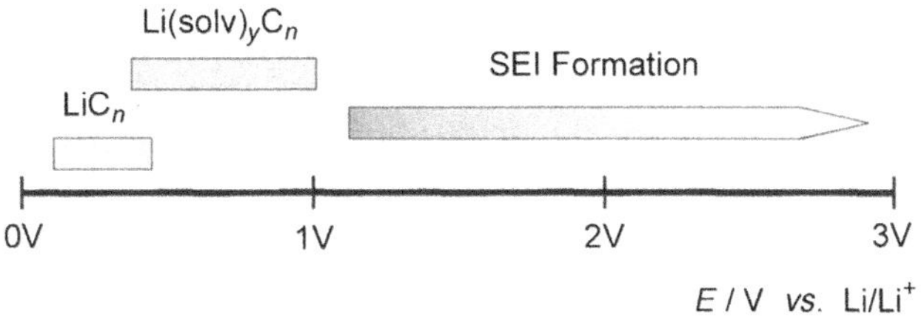

Fig. 4. Approximate potential regions for unsolvated lithium intercalation (LiC_n) and solvated lithium intercalation [$Li_x(solv)_yC_n$]; the potential region where electrolyte reduction and subsequent SEI formation should take place has to be higher than the potential region for solvated lithium intercalation

Table 1. Specific charges for charge (C_c) and discharge (C_d) as well as discharge/charge efficiencies (*Eff* = ratio of discharge capacity to charge capacity) of TIMREX® graphite SFG 44 in (a) 1 *M* $LiClO_4$ in *PC/ES* (95/5), (b) 1 *M* $LiClO_4$ in *PC/PS* (95/5), (c) 1 *M* $LiClO_4$ in *PC/DMS* (95/5), (d) 1 *M* $LiClO_4$ In *PC/DES* (95/5), (e) 1 *M* $LiClO_4$ in *PC/ES* (97/3), and (f) 1 *M* $LiClO_4$ in *PC/ES* (99.5/0.5); constant current charge/discharge cycling with $i = \pm 0.02\,A \cdot g^{-1}$, cut-off: 1.8/0.025 V *vs.* Li/Li$^+$

	Cycle Number	C_c / $Ah \cdot kg^{-1}$	C_d / $Ah \cdot kg^{-1}$	*Eff* / %		Cycle Number	C_c / $Ah \cdot kg^{-1}$	C_d / $Ah \cdot kg^{-1}$	*Eff* / %
(a)	1	429	342	79.7	(b)	1	485	338	69.7
	2	362	347	95.8		2	396	344	86.7
	3	359	350	97.7		3	389	355	91.3
	4	357	351	98.3		4	381	353	92.5
	5	353	349	98.9		5	374	355	94.9
(c)	1	555	338	61.0	(d)	1	689	301	43.7
	2	396	348	87.9		2	380	315	82.7
	3	372	342	91.9		3	356	316	88.7
	4	362	336	92.8		4	354	311	88.0
	5	364	335	92.1		5	372	305	81.9
(e)	1	425	332	78.1	(f)	1	427.5	317.3	74.2
	2	359	339	94.4		2	374.1	324.0	86.6
	3	359	341	94.9		3	364.1	324.0	89.0
	4	357	342	95.8		4	369.1	329.0	89.1
	5	352	342	97.2		5	364.1	329.0	90.5

i.e. ES is the most effective electrolyte additive (Table 1a–d). Even concentrations of *ES* as low as 0.5% (v/v) still enable successful cycling of graphite anodes in *PC*-based electrolytes. However, we found that the *ES* additive amount has to be in the range of 3–5% (v/v) in order to keep the discharge/charge efficiencies sufficiently high, *i.e.* to keep C_{irr} low (Table 1a,e,f).

Fluorinated solvents for lithium ion battery electrolytes

The redox potential of lithium-rich LiC_n compounds (100 mV *vs.* Li/Li$^+$) is rather close to that of metallic lithium, and the particle size of LiC_n in battery electrodes is typically in the order of ~10–40 µm, which means that the reactive surface area with the electrolyte is large. The kinetic stability against a spontaneous reaction with the electrolyte is determined by the protective SEI [5]. In order to reach an optimum electrolyte conductivity, the electrolytes usually contain up to 66% (v/v) of the low viscosity solvents *DMC* (boiling point: 90°C, flash point: 18°C) and/or *DEC* (boiling point: 127°C, flash point: 31°C)[3]. When the protective interphases collapse, these volatile and highly flammable low viscosity components may readily react with the anode and contribute to a rapid thermal runaway of the cell[4]. A replacement of these components in future cells therefore seems inevitable.

Many partly fluorinated polar organic solvents such as ethers, esters, amides, and others show considerable solubilities for lithium salts, at least for those with large anions such as PF_6^- and $[N(SO_2CF_3)_2]^-$ [13–18], despite the strong electron-withdrawing effect of the fluorine atoms on the electron donor groups (usually containing oxygen) which are required for Li^+-ion solvation. Exchange of some hydrogen for fluorine causes significant differences in the physical and chemical properties of polar solvents. In particular, viscosities, but also melting and boiling points of fluorinated solvents, are in many cases significantly lower compared to their hydrogenated counterparts [13, 14]. Moreover, they are in general much less flammable because less hydrogen is available.

Typical examples of fluorinated organic solvents which have been investigated are glycol ethers of the type $HC_2F_4O(C_2H_4O)_nC_2F_4H$ $n=1, 2, 3, 4, 6$ [15]), urethanes [16] such as $(CH_3)_2NCO_2CH_2CF_3$, glycol esters [17], and acetamides [18] as well as selected fluorinated carbonates, acetates, ethers, borates, and phosphates.

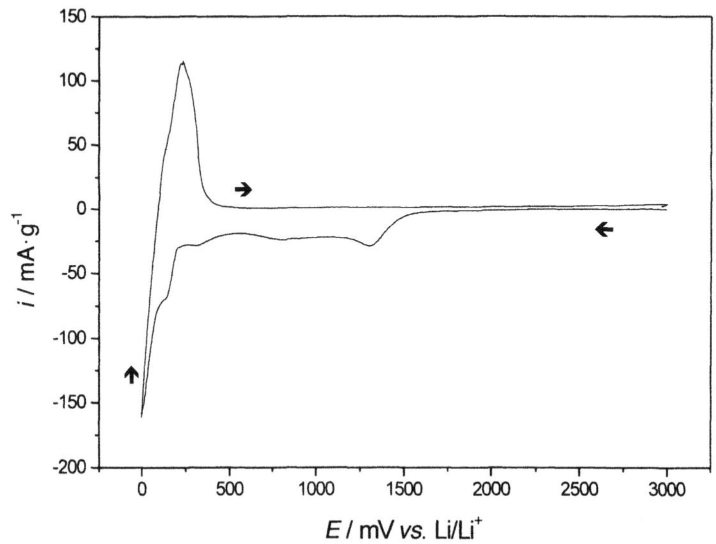

Fig. 5. Cyclic voltammogram of synthetic graphite TIMREX® KS6 in 1 *M* $LiN(SO_2CF_3)_2$ in *PC/BTE* (90/10); scan rate: 0.1 $mV \cdot s^{-1}$

[3] By contrast, *PC* and *EC* show much higher boiling and flash points (*PC*: boiling point: 240°C, flash point: 132°C; *EC*: boiling point: 244°C, flash point: 160°C).

[4] In addition, also the safety behavior at the positive electrode depends on the formation of a protective interphase and furthermore on the state of charge and the temperature conditions [4].

Table 2. Specific charges for charge (C_c) and discharge (C_d) as well as discharge/charge efficiencies (*Eff* = ratio of discharge capacity to charge capacity) of TIMREX® graphite SFG 44 in 1 *M* $LiN(SO_2CF_3)_2$ in *PC/BTE* (80/20); constant current charge/discharge cycling with $i = \pm 0.02\,A \cdot g^{-1}$, cut-off: 1.0/0.024 V *vs.* Li/Li^+

Cycle Number	$\frac{C_c}{Ah \cdot kg^{-1}}$	$\frac{C_d}{Ah \cdot kg^{-1}}$	$\frac{Eff}{\%}$
1	358	277	77.2
2	295	285	96.6
3	291	286	98.4
4	288	285	99.0
5	296	293	99.1
6	283	282	99.7
7	287	286	99.7
8	287	287	99.7
9	283	283	99.8
10	285	285	100.0

As an example we here present data on 1,2-*bis*-(trifluoracetoxy)-ethane (*BTE* [17]; boiling point: 157°C, flash point >110°C). Like *ES* (Fig. 3b), *BTE* renders possible the operation of graphitic anodes in *PC* based electrolytes (Fig. 5). An amount as low as 10% (v/v) of the fluorinated compound in the electrolyte sufficiently suppresses *PC* co-intercalation into graphite. The SEI formation due to reduction of *BTE* starts at 1.75 V *vs.* Li/Li^+, *i.e.* at potentials higher than the potential of solvated intercalation, and thus allows reversible lithium intercalation. These observations agree with the results of the constant current charge/discharge experiments summarized in Table 2.

SEI formation on lithium storage metals and alloys

In order to overcome the problems with the mechanical disintegration of lithium storage metals and alloys during cycling (see above), several simple strategies for the design of the metallic host materials have to be followed. These strategies are summarized in the following (for more details, see Refs. [6, 19]).

(a) Proper design of particle morphology, especially particle size and porosity

(See *e.g.* Refs. [19–21]) A nano-structured material shows the same relative expansion as a polycrystalline material, but the absolute expansions are lower, and hence the cycling stability increases. Pores in the original material should offer additional space for expansion and therefore also increase the cycling stability.

(b) Incorporation of inactive components

The dilution of the active material with matrix components results in an increased cycling stability, since the reacting and expanding phase is stabilized by the non-reacting inactive matrix. This concept has first been described in detail by *Huggins et al.* [22] in the mixed-conductor matrix concept and also holds true for a number of recently proposed promising materials, such as *Fuji*'s amorphous tin-based composite oxide [23] and tin-oxides in general (*e.g.* Ref. [24]) or materials from the Sn-Fe(-C) system [25], to mention just a few. The drawback of these active/inactive composites [25] is that the presence of inactive matrix components reduces the specific charge as well as the charge density of the electrode material.

(c) Use of multi-phase instead of single-phase materials

In multi-phase materials like *e.g.* Sn/SnSb[5] or Sn/$SnAg_3$ [19–21], the single components usually react one at a time. Hence, while one component is reacting (expanding or contracting), the other(s) do(es) not and can thus stabilize the reacting phase.

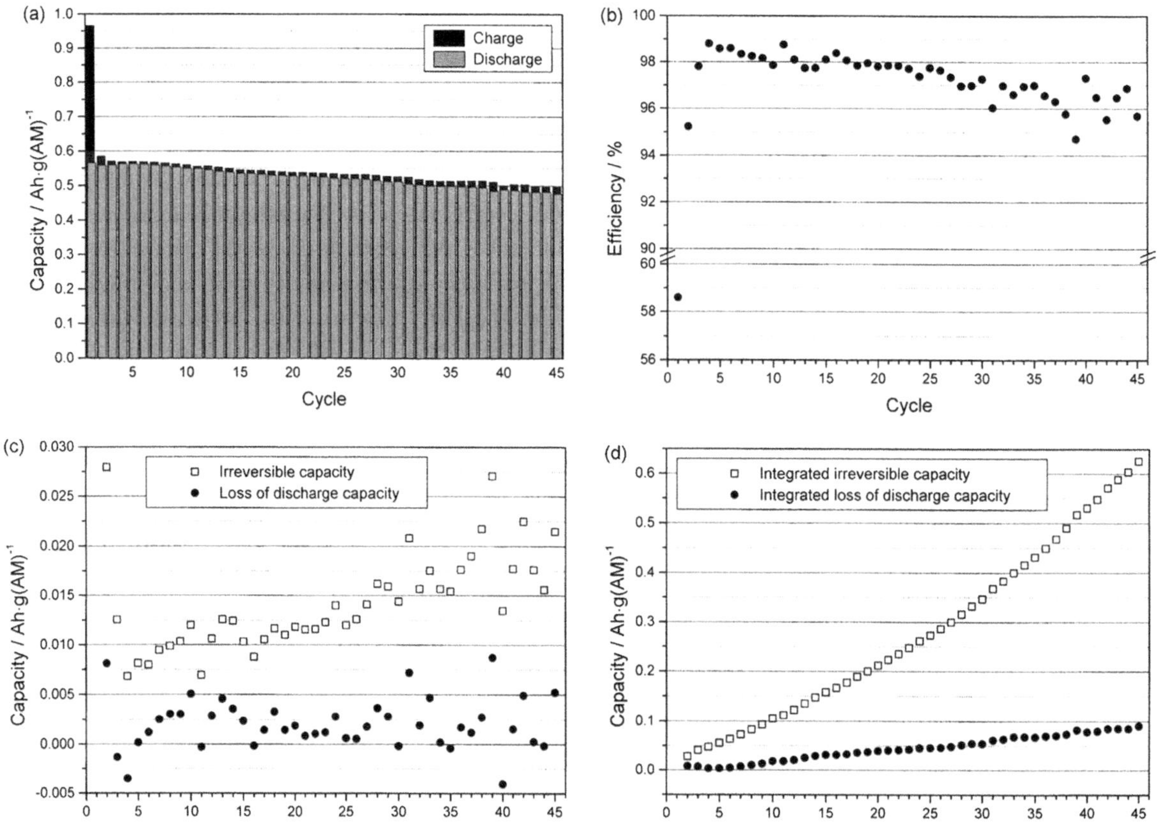

Fig. 6. Cycling performance of a Sn/SnSb composite electrode; (a) charge and discharge capacities (AM = active material), (b) efficiencies, (c) irreversible capacity and loss of discharge capacity (LDC) per cycle, (d) integrated irreversible capacity and integrated LDC (beginning with the 2nd cycle); electrolyte: 1 *M* $LiClO_4$ in *PC*; constant current charge/discharge cycling with $i = \pm 0.3 mA \cdot cm^{-2}$) (approx. $0.044 A \cdot g^{-1}$), cut-off: 1.2/0.1 V *vs.* Li/Li^+

[5] Cf. Ref. [19] for problems with the stoichiometry and structure of "SnSb".

(d) Use of components such as SnSb which show reversible phase separation and restoration upon reaction with Li [19–21]

XRD investigations of SnSb at various stages during charge point to a reaction mechanism where Li_3Sb (or $Li_{2+x}Sn_ySb$) is formed and in parallel Sn is segregated (which subsequently reacts reversibly with Li). This means that additional nano-structuring occurs *in situ*. Furthermore, this mechanism is reversible, which should at least to some extent counteract the aggregation of Sn into larger clusters during cycling (giving rise to problems with cracking and crumbling dur to the volume changes), as was found by *Courtney et al.* [26] for Sn-oxide based systems.

By following the above mentioned strategies – except point (*b*), which was avoided, since it produces additional weight and volume and thus reduces the gravimetric and volumetric lithium storage capacities of the material – it was possible to develop a composite electrode containing a nano-structured Sn/SnSb powder which shows reversible capacities of more than $500\,mAh \cdot g^{-1}$ (with respect to the mass of active material) for more than 30 cycles (Fig. 6a).

From the cycling results (Figs. 6a and b) it is evident that – apart from a further improvement of the cycling stability – future investigations will have to focus on the irreversible capacities which are high, especially in the first, but also exist in the following cycles. Some major reasons for the irreversible capacity are listed below.

(*i*) Reduction of electrolyte solution and formation of the SEI as has been discussed above for carbonaceous anodes.

(*ii*) Reduction of oxide impurities. Due to the preparation method (see Experimental), the Sn/SnSb as well as the Ni(B) powder (the latter is used as conductive electrode additive) contain oxygen impurities which will be irreversibly reduced during the first lithium uptake. This reaction should mainly occur in the first cycle.

(*iii*) Loss of contact inside the active material due to cracking and crumbling. Hence, less and less active material can partake in the charge/discharge reactions, *i.e.* the uptake and release of Li.

(*iv*) Trapping of Li in the active material, *i.e.* not all of the Li which was inserted during charge can be extracted again during discharge at the applied currents and potentials. In the case of lithium storage metals and alloys it has been frequently observed that in the cycles following the very first one the efficiencies exceed 100%. This is explained with a change in morphology (further nano-structuring during cycling, increase of porosity, *etc.*) which leads to improved kinetics (*e.g.* shorter diffusion pathways). Thus, Li which has been trapped in the initial cycle(s) may be extracted in later ones [27].

A rough estimation of the importance of the various effects is possible by comparing the irreversible capacity (C_{irr}), calculated as the difference between charge and discharge capacity, with the loss of discharge capacity (LDC) defined as the difference between the discharge capacities of two successive cycles. Since no LDC can be given for the first cycle and since the situation for the first cycle is a special one (oxide reduction, first SEI formation, first expansion), only results from cycle 2 upwards are compared in Fig. 6c (absolute values) and Fig. 6d (integrated values, beginning with the 2nd cycle). Whereas all four mentioned effects are responsible for C_{irr}, in a first approximation only loss of inter-particle contact (effect (*iii*)) and Li trapping (effect (*iv*)) should contribute to the LDC[6]. The C_{irr} values integrated from cycle 2–45 are several times larger than the integrated LDC values (Fig. 6d). Except for cycle 1, where significant surface oxide reduction can be anticipated, SEI formation is the dominating effect to C_{irr} of the metal host anodes and obviously does not diminish in the later charge/discharge cycles as in the case of graphitic anodes (cf. Table 2)[7].

Conclusions

If solvated intercalation is suppressed, graphitic carbons are dimensionally quite stable during cycling (only about 10% volume change during Li intercalation/de-intercalation). Electrolyte reduction is restricted to the first few cycles until an effective (pin-hole free, electronically insulating, Li^+-cation conducting, and dimensionally stable during cycling) SEI is formed (Fig. 7a).

By contrast, the metal anode/electrolyte interface is quite variable during cycling, due to the large volume changes. The once formed film may break up, and parts of it may even break off. Whenever fresh, unfilmed surface of the active material is exposed to the electrolyte, a new SEI will be formed (Fig. 7b). Hence, filming will extend over a larger number of cycles, maybe all throughout cycling. The ideal SEI for lithium storage metals and alloys is therefore a thin one, which is associated with a minimum of lithium and electrolyte loss, and a flexible one, which can better withstand the volume changes during cycling. Though the reversible capacities (discharge capacities) and the cycling stability of the metal host anodes

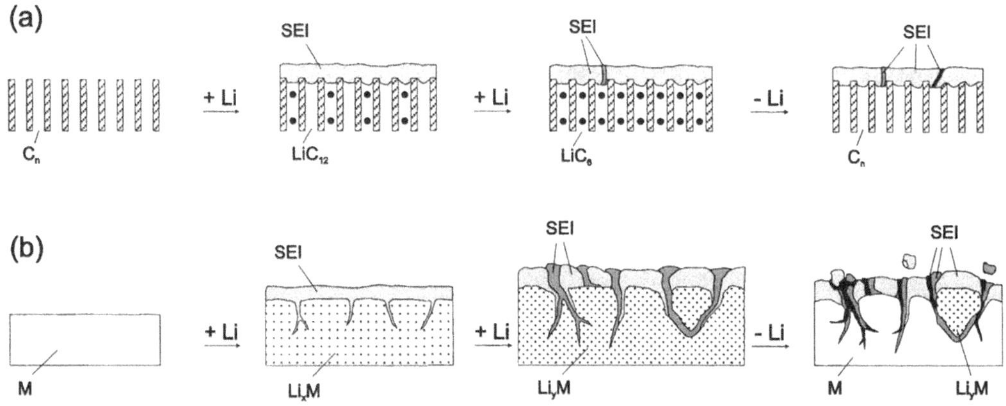

Fig. 7. Model for SEI formation on (a) carbonaceous and (b) metallic host materials; different shading of the SEI indicate SEI formed at different stages of Li uptake and release (not different composition)

[6] It should be noted that this model is simplified, as it does not consider, for instance, effects caused by the changing morphology (which influences the electrode kinetics), by the increasing electrode impedance (which results from film growth), or a break-up of the SEI during discharge which is followed by new film formation (at the expense of Li, which is extracted from the metal/intermetallic), *etc*.

[7] For the present electrode and cell design and cycling mode.

are quite good, both the electrolyte composition and the metal anode surface have not been optimized with respect to adequate SEI formation. This task is subject to present investigations.

Experimental

Synthesis of lithium storage alloys

The nanocrystalline Sn/SnSb and Ni(B) powders were precipitated with $NaBH_4$ from aqueous solutions of the respective chlorides (the former in the presence of complexants). The particle size of the Sn/SnSb powder was below 300 nm, the *BET* surface area was $14.5\,m^2\,g^{-1}$, the nominal composition was $Sn_{0.88}Sb_{0.12}$, and the Ni(B) powder contained approximately 8% (w/w) B. Further details on the synthesis of the powders can be found elsewhere [19].

Electrolytes, electrodes, and cells

ES (Aldrich, 98%) was distilled under vacuum. *EC*, *PC*, *DMC*, and $LiClO_4$ (all Merck, battery grade) were used as received. $LiN(SO_2CF_3)_2$ (3 *M*) was dried under dynamic vacuum at 110°C for 48 h. *BTE* was synthesized and purified as described in Ref. [18]. All electrolyte components (solvents and electrolyte salts) were handled by standard procedures (*e.g.* Refs. [8, 13]). The ratios of the solvents in the electrolyte mixtures are given in volume percent. The water content of the electrolytes was typically less than 15 ppm as determined by *Karl Fischer* titration. Graphite based composite anodes were made from TIMREX® graphites and 5% (w/w) polyvinylidene fluoride (*PVDF*) binder as described elsewhere [8]. Composite lithium storage alloy electrodes were prepared by pasting a well-mixed slurry of 82% (w/w) Sn/SnSb, 10% Ni(B), and 8% *PVDF* onto stainless steel mesh, pre-drying, pressing, and final drying in vacuum at 120–140°C (for more details, see Ref. [19]). Electrolyte preparation and cell assembly were accomplished under a dry argon atmosphere in a glove box. Electrochemical experiments were carried out in laboratory type glass cells with bulk lithium counter and reference electrodes and an excess of electrolyte. The electrodes were not closely packed in separator materials but placed in the electrolyte without any further support or protection. The constant current charge/discharge and voltammetric experiments were performed with ADESYS electrochemical testing units developed in our laboratory [28].

Acknowledgements

Financial support by the *Austrian Science Foundation* and the *OeNB* within the special research program *Electroactive Materials* and Project No. FWF-P12768-CHE is gratefully acknowledged. The work on lithium storage alloys is supported by Mitsubishi Chemical Corp. (Japan). Merck KGaA (Germany) and the TIMCAL group (Switzerland) have provided samples used in this study.

References

[1] Nagaura T, Tozawa K (1990) Prog Batt Solar Cells **9**: 209
[2] Nomura Research Institute (1996) Nomura Institute Research Report, Advanced Rechargeable Battery Industry '96, Tokyo, Japan
[3] See for example papers in: ITE Batt Lett (1999) Vol 1(2)
[4] Winter M, Besenhard JO, Spahr ME, Novák P (1998) Adv Mater **10**: 725
[5] Winter M, Besenhard JO (1999) Lithiated Carbons. In: Besenhard JO (ed) Handbook of Battery Materials, part III. Wiley-VCH, Weinheim, p 383
[6] Winter M, Besenhard JO (1999) Electrochim Acta **45**: 31

[7] Peled E, Golodnitzky D, Penciner J (1999) The Anode/Electrolyte Interface. In: Besenhard JO (ed) Handbook of Battery Materials, part III. Wiley-VCH, Weinheim, p 419
[8] Wrodnigg GH, Besenhard JO, Winter M (1999) J Electrochem Soc **146**: 470
[9] Besenhard JO, Wagner MW, Winter M, Jannakoudakis AD, Jannakoudakis PD, Theodoridou E (1993) J Power Sources **43–44**: 413
[10] Imhof R, Novák P (1996) J Electrochem Soc **145**: 1081
[11] Wrodnigg GH, Wrodnigg TM, Besenhard JO, Winter M (1999) Electrochem Comm **1**: 148
[12] Wrodnigg GH, Reisinger C, Besenhard JO, Winter M (1999) ITE Batt Lett **1**: 110
[13] Lie LH, Hodal T, Möller KC, Wrodnigg GH, Appel WK, Besenhard JO, Winter M (1999) ITE Batt Lett **1**: 106
[14] Möller KC, Hodal T, Appel WK, Besenhard JO, Winter M (to be published)
[15] Besenhard JO, v. Werner K, Winter M (1997) Ger Pat PCT/DE97/107506.4-1215 Eur Pat 807986, US Pat 5916708
[16] Appel WK, Besenhard JO, Pasenok S, Winter M, Wrodnigg GH (1998) Ger Offen DE 19 724 709 Al
[17] Appel WK, Besenhard JO, Lie LH, Pasenok S, Winter M (1998) Eur Patent Appl
[18] Appel WK, Besenhard JO, Lie LH, Pasenok S, Winter M (1998) Eur Patent Appl
[19] Yang J, Wachtler M, Winter M, Besenhard JO (1999) Electrochem Solid-State Lett **2**: 161
[20] Besenhard JO, Yang J, Winter M (1997) J Power Sources **68**: 87
[21] Yang J, Winter M, Besenhard JO (1996) Solid State Ionics **90**: 281
[22] Boukamp BA, Lesh GC, Huggins RA (1981) J Electrochem Soc **128**: 725
[23] Idota Y, Kubota T, Matsufuji A, Maekawa Y, Miyasaka T (1997) Science **276**: 1395
[24] Courtney IA, Dahn JR (1997) J Electrochem Soc **144**: 2045
[25] Mao O, Turner RL, Courtney IA, Fredericksen BD, Buckett MI, Krause LJ, Dahn JR (1999) Electrochem Solid-State Lett **2**: 3
[26] Courtney IA, McKinnon WR, Dahn JR (1999) J Electrochem Soc **146**: 59
[27] Yang J, Besenhard JO, Winter M (1997) In: Holmes CF, Landgrebe AR (eds) The Electrochemical Society Proceedings Series 97-18, Batteries for portable applications and electric vehicles, p 350
[28] Evers B, Schneider I, Wrodnigg GH, Winter M, Besenhard JO (March 31, 1998) In: Schwabe I (ed) Begleittexte zum Entwicklerforum: Batterien, Ladekonzepte & Stromversorgungsdesign. München, p 28

Received May 30, 2000. Accepted June 14, 2000

A Method of Manufacturing Highly Conductive Composite Materials by Coating Surfaces of Nonconductors with Fine Particulate Conductive Substances

Susanne Voß[1,*], **Vadim Nigmatoulline**[1], **Marjan Bele**[2], **Stane Pejovnik**[2], and **Jürgen O. Besenhard**[1]

[1] Institute for Chemical Technology of Inorganic Materials, Graz University of Technology, A-8010 Graz, Austria

[2] National Institute of Chemistry, SL-1115 Ljubljana, Slovenia

Summary. A widely applicable aqueous dip-coating process termed substrate induced coagulation (SIC) suitable for coating various types of substrates (*e.g.* polar and unpolar polymers, inorganic glasses, metals) with fine particulate materials (*e.g.* carbon blacks, small particle size SiO_2 or TiO_2) has been developed. This process is based on the interaction of a conditioner (water soluble polymers or polyelectrolytes, *e.g.* proteins, polyvinylalcohols, polyacrylates with low charge density) which is adsorbed on the substrate with a surfactant-stabilized dispersion containing some additional electrolyte. Employing dispersions of conductive particles, the resulting coatings can be used as a conductive starting layer for electroplating. Such layers are fairly rinse-proof and may be applied in the metallization process of through holes in printed wiring boards or for the production of highly conductive composite materials from coated particles. SIC has also been used to improve the distribution of carbon in composite battery electrodes such as MnO_2/C.

Keywords. Conductive composite materials; Improved percolation pattern; Carbon black; Substrate induced coagulation; Dip coating.

Introduction

The production of conductive composites from predominantly non-conductive materials by adding conductive substances is attractive for many areas of application, such as conducting, and accordingly shielding, plastic housings for electronic devices or lightweight electrodes and current collectors for electrochemical applications. Poorly conductive active masses of battery electrodes, such as manganese dioxide, also require more conductive additives (usually carbons), and the performance of the battery electrodes depends heavily not only on the kind and amount of these additives, but also on their spatial distribution.

---

* Corresponding author

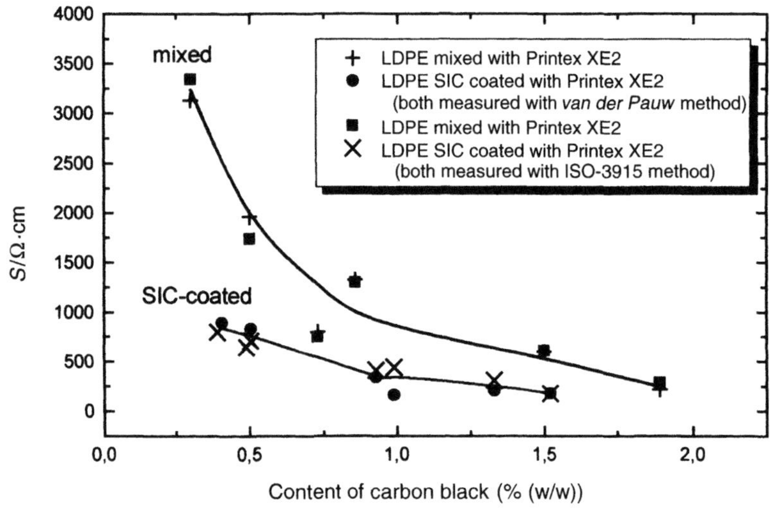

Fig. 1. Dependence of the resistivity on the amount of conductive carbon black Printex XE2

It has been demonstrated in many studies [1–10] that the increase in the conductivity of composites is by no means proportional to the addition of the conductive component. Only after exceeding a threshold concentration of the conductive component(s), which is usually quite high, a significant increase in conductivity is observed. Substantial conductivity of the composite only occurs when the conductive component is so closely packed that chains and networks of conductive particles can be formed. Assuming uniformly shaped particles, the conductivity of mixtures of conductive and non-conductive particles can be calculated by the percolation theory [11]. The obvious advantage of SIC-coating over simple mixing, in particular at low content of carbon black, is illustrated in Fig. 1.

The objective of our work was to provide a method for the fabrication of highly conductive composites containing only small amounts of the conductive component by creating favourable distribution patterns of the same. The basic principle of the method is that the non-conductive particles are *i*) uniformly coated with a layer of the conductive component and then *ii*) compacted, *e.g.* by pressure molding. This creates a 3-dimensional network of the conductive component which cannot be achieved by simple mixing. 3-Dimensional metallic networks can be easily created by metallizing the non-conductive component by means of electroless (= chemical) metallization processes [12–14]. In fact, highly conductive composites can be prepared from non-conductive particles coated with extremely small amounts of the conductive component [12]. On the other hand, the further processing of the coated polymer particles requires some precautions *i*) to avoid disintegration of the well-ordered distribution of the conductive component and *ii*) to make sure that the mechanical properties of the composites are still satisfactory. In this paper we report on the preparation of carbon black/polymer composites. The uniform coating of the polymer particles is done by SIC. This method has been developed originally as a dip-coating process, in particular for creating conductive layers in the holes of double-sided or multi-layer printed wiring boards [15–18]. In this kind of operation,

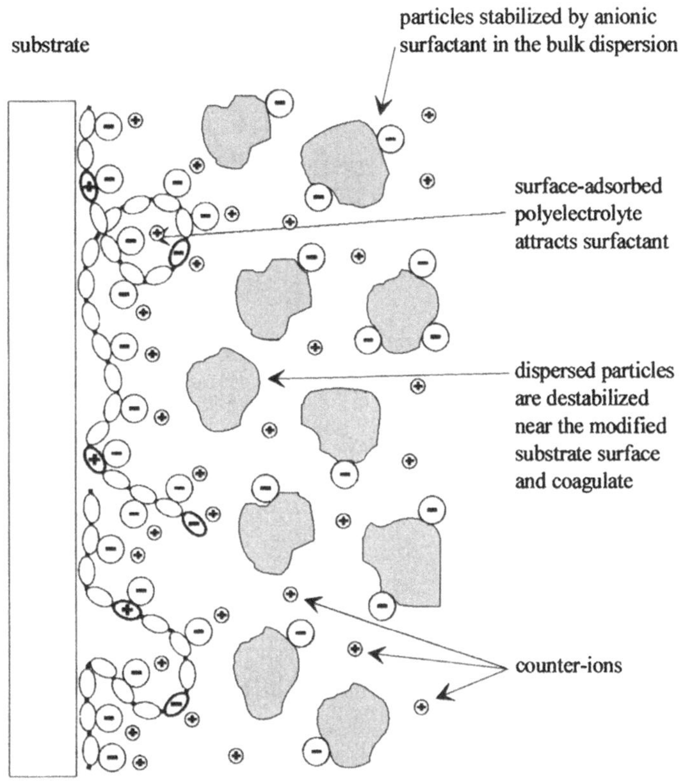

Fig. 2. Principle of the coagulation procedure at a conditioned surface

the concentrations of the active components in the solutions used for conditioning and coating are only slowly decreased and may be readjusted continuously by an automatic system [15].

A modified discontinuous batch version of the SIC process can also be used to create uniform coatings on larger quantities of powder substrates exhibiting large surface areas and, hence, changing drastically the concentrations of the solutions used for treatment.

Attraction of the surfactants by the adsorbed conditioner may be the main driving force for the SIC process (see Fig. 2). There are also indications for a direct interaction between the polymer conditioner and the particles in the dispersion [19–22].

Results and Discussion

The conductivity of carbon black polymer composites always increases in a strongly non-linear manner with the amount of carbon black [10, 25, 26]. The critical level of percolation of conventionally prepared carbon-polymer composites (~10%) can be significantly decreased by using SIC carbon coated polymer particles for compression molding of the electrode. The thickness of carbon layers created by the SIC dip-coating process is typically *ca.* 200–500 nm [18, 27, 28]. Thicker layers can be easily obtained by multiple application of the SIC process. To get a uniform and rinse-proof deposition, the parameters of the SIC process have to be adjusted with respect to the substrate and the kind of particles to be coated onto that substrate.

Optimization of the SIC process requires consideration of *i*) the kind of conditioner, *ii*) the kind and concentration of the surfactant stabilizing the dispersion, and *iii*) the kind and concentration of the electrolyte in that dispersion and, of course, of various process parameter such as temperature, duration of treatments, *etc*.

Effect of conditioner

Various kinds of water-soluble polymers, polyelectrolytes as well as uncharged polymers, may be used as conditioners. Surprisingly, positively and negatively charged polyelectrolytes are effective in combination with positively as well as negatively charged surfactants stabilizing the dispersed particles, *i.e.* the SIC process is not based on a simple electrostatic interaction between conditioner and surfactant. Figure 3 illustrates the effect of different gelatine conditioners on the carbon black deposition from a dispersion containing Printex L6, AOT, and NaOAc onto epoxy-based printed wiring board material (FR-4 Isola Düren, Germany). The *pH* of the conditioner solutions was adjusted to the IEP of the respective gelatin. A-Type gelatines are obtained by acidic hydrolysis of collagens, B-type gelatines by basic hydrolysis. The *Bloom* value is a measure for the gel strength and usually increases with molecular mass and degree of crosslinking of the gelatine. Results obtained with polyvinylalcohols (Mowiol®, Hoechst, Germany) were comparable to those with low gel strength gelatins. On the other hand, carboxymethylcellulose-Na-salt turned out to be ineffective as a conditioner; obviously, the charge on this polymer is too high for being strongly adsorbed on the substrate.

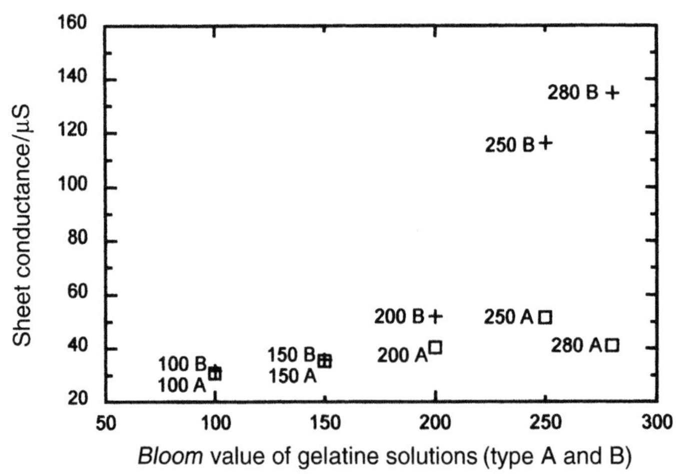

Fig. 3. Sheet conductance of carbon black coated printed wiring boards using different kinds of gelatine as conditioners

The surface conductivity of the carbon black coated samples (determined by the *van der Pauw* method) was used as a quick and simple measure for the effectiveness of coating. The results indicate that conditioning with high gel strength B-type gelatines leads to high sheet conductivities.

Effect of concentration of surfactant in the carbon black dispersion

Several authors [26, 29–33] have studied the adsorption of surfactants and, in particular, of *CTAB* [29–31] onto carbon black in aqueous solutions where the dispersion of carbon black particles is maintained by a repulsive barrier of surface charges or steric interaction of the adsorbed surfactant. *Langmuir*-type adsorption isotherms with long plateau regions are found for most of the carbon blacks.

The amount of surfactant present in the carbon black dispersion is an extremely important parameter, because an excess of surfactant deactivates the conditioner layer adsorbed on the substrate. Fortunately, there is a strong interaction between various surfactant molecules and carbon black, and the point where significant amounts of free surfactant (not adsorbed on carbon black) show up in solution can be determined easily, *e.g.* by polyelectrolyte titration in the supernatant (see Fig. 4).

Polyelectrolyte titration shows, in agreement with earlier reports [31], a linear dependence of the adsorbed amount of surfactant on the equilibrium concentration in the solution. The saturation plateau value of *CTAB* of 1.8 $mmol \cdot g^{-1}$ represents the maximum amount of *CTAB* surfactant adsorbed per mass unit of Printex XE2.

The amount of surfactant adsorbed on carbon black can also be estimated by X-ray diffraction, assuming that diffraction peaks corresponding to crystalline surfactants will not show up for adsorbed surfactant. The X-ray patterns of pure Printex XE2 and air-dried dispersions are presented in Fig. 5. Strong diffraction peaks of crystalline *CTAB* show up from 2 mmol of *CTAB* per g of Printex XE2 upwards. These results agree well with those from polyelectrolyte titration. From FTIR data it has been concluded that the CTA^+ cation is adsorbed on the carbon black particles

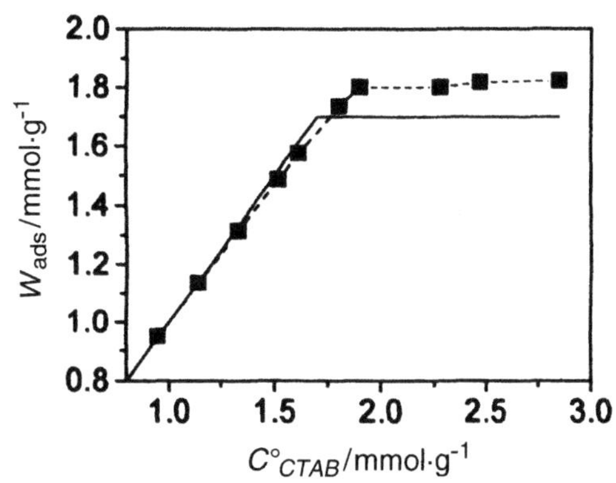

Fig. 4. Adsorption isotherm of *CTAB* adsorbed on Printex XE2 at room temperature; the initial concentration refers to the mass of Printex XE2; the bold line schematically shows the adsorbed amount of *CTAB* in an air-dried dispersion as deduced from X-ray data with the assumption that all *CTAB* in the air-dried dispersion is adsorbed in a sample where *CTAB* diffraction peaks are absent (see Fig. 5)

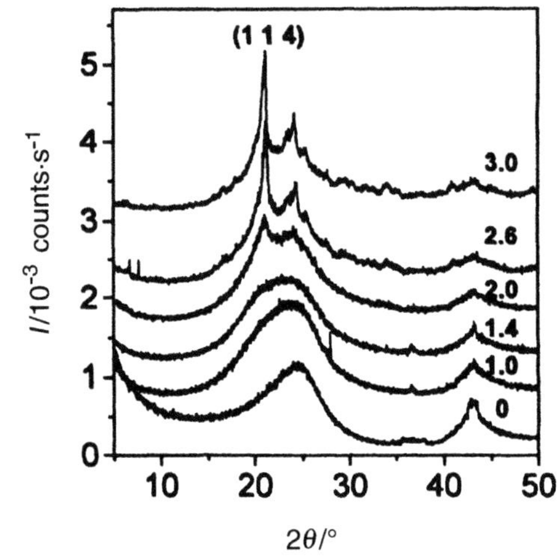

Fig. 5. X-Ray diffraction data from air-dried powders of carbon black dispersions; the amount of *CTAB* per mass unit of Printex XE2 ($mmol \cdot g^{-1}$) is used as the label

by the combined action of the hydrophobic interaction between the non-polar part of the CTA^+ cation and the hydrophobic parts of the surface of carbon black particles [28].

Composites for carbon coated polymer powders

Coating of polymer powders by the SIC process was done as reported earlier in detail [17, 33–36]. After coating, the dried powders were compacted by compression molding at various periods of time, temperatures, and pressures. The processing temperature during molding turned out to be the most important parameter for the optimization of the conductivity and also of the mechanical properties of the composites. If the molding temperature (or pressure) is too low, the mechanical stability of the composites will be insufficient. On the other hand, processing temperatures close or above the melting point will destroy the 3-D percolation pattern of the carbon black and result in drastically decreased conductivity. For this reason, processing parameters have to be kept in a fairly narrow window. This is illustrated by the values of the modulus of elasticity and the resistivity presented in Tables 1 and 2 for polyolefins containing *ca.* 0.1% (w/w) of carbon black Printex XE2. Obviously, the granules are deformed at a temperature of about 160–162°C, but the original particle size is still recognizable after the molding process. Only a temperature around 164°C allows the production of a molded pellet with a homogeneous distribution. This is reflected in the values of the resistivities given in Table 1.

At temperatures above the melting point (164.8°C) of polypropylene the resistivity increases rapidly, and no homogeneous percolation pattern is be obtained. As the processing temperature of the thermoplastic non-conductors is very low, composites of polymers with different melting and softening point should lead to better results.

Table 1. Properties of untreated and carbon black (Printex XE2) coated polymers depending on processing temperature

Polymer	Processing temperature (°C)	Properties	
		Modulus of elasticity ($N \cdot mm^2$)	Resistivity ($\Omega \cdot cm$)
Polypropylene (PP)		954	$>10^{17}$
PP coated	164	–	290
with Printex XE2	166	–	7944
LDPE	130	193	$>10^{17}$
LDPE coated	100	204	56
with Printex	105	240	130
XE2	130	219	265
HDPE 5021D	160	872	$>10^{17}$
HDPE 5021D	120	–	56
coated with	128	878	233
Printex XE2	160	876	380
HDPE 5060D	160	801	$>10^{17}$
HDPE 5060D	120	–	146
coated with	125	972	1425
Printex XE2	160	812	3103

Table 2. Mechanical properties of carbon black coated polymers (0.1% (w/w) carbon black) measured as recommended in DIN 53455 [37] and DIN 53457 [38]

Mechanical properties	Polymer composites		
	LDPE 1800 with 0.1% (w/w) Printex XE2	HDPE 5021D with 0.1% (w/w) Printex XE2	HDPE 5060D with 0.1% (w/w) Printex XE2
Tensile stress at yield in MPa	9.05	23.9	24.8
Tensile stress at offset yield point in %	20	12	12
Tensile stress at break point in MPa	8.83	18.9	17.7
Tensile stress at offset break point in %	297	527	50

For mixtures of carbon black coated PP and HDPE the processing windows turned out to be significantly larger (see Fig. 5). Conductivity measurements of PE and PP samples (low-density polyethylene LDPE 1800 SP15, BASF, Germany; high-density polyethylene HDPE Lupolen 5021 D and Lupolen 5060 D, BASF, Germany; polypropylene PP-PCD, Linz, Austria) coated with carbon black are presented in Figs. 6 and 7. Conductivities were usually measured by the *van der Pauw* method. Bulk conductivities measured between polished contact electrodes showed similar results but were less reproducible.

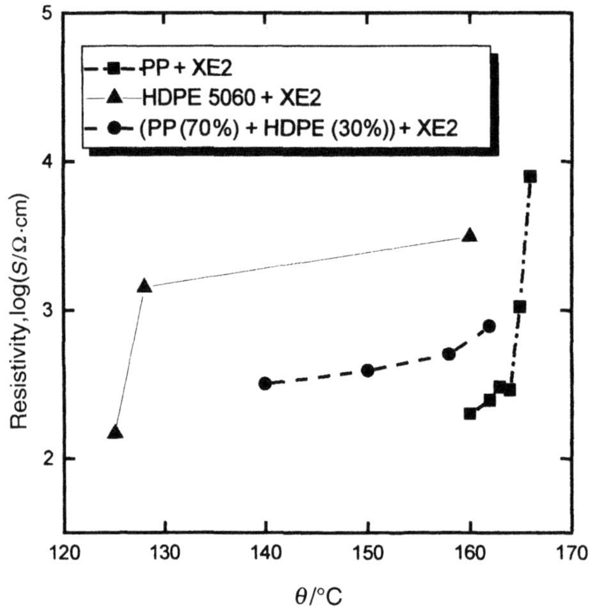

Fig. 6. Dependency of the resistivity on the molding temperature (θ) for carbon black coated polymers; the content of carbon black Printex XE2 is in the order of 0.1% (w/w)

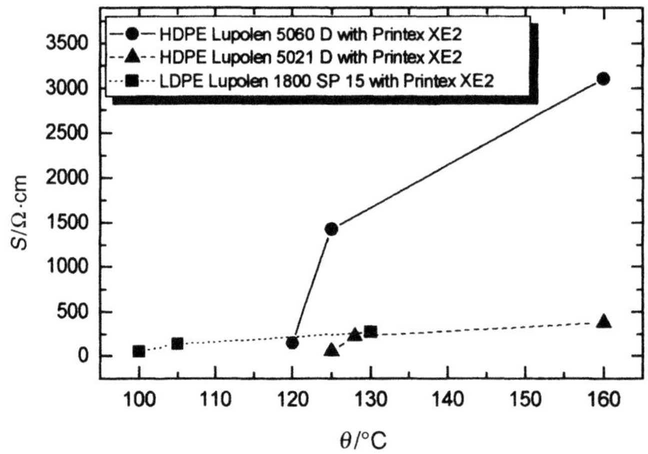

Fig. 7. Resistivity of carbon black coated polymers; the content of carbon black Printex XE2 is in the order of 0.1% (w/w)

Conclusions

The production of conductive composites by compression molding of polymer particles coated with carbon black by substrate induced coagulation is an attractive method to achieve high conductivities at low content of carbon black. Unfortunately, the processing windows with respect to molding temperature and pressure are fairly narrow, in particular for single-component polymers. Processing of polymer blends (hard/soft or lower/higher melting point) turned out to be much easier.

Experimental

The general procedure as reported in Ref. [17] and illustrated by Fig. 2 was used throughout this study. Highly conductive carbon black, either Printex L6 or Printex XE2 (Degussa, Germany) were coated onto polyethylene and polypropylene (HDPE, LDPE, and PP, BASF, Germany) after conditioning with different types of coagulation inducers depending on the kind of substrate. In particular, bath compositions and process parameters were as follows: The aqueous 0.2% (w/w) polymer solution contained either gelatine (type A or B with different *Bloom* values, Fluka or Deutsche Gelatine-Fabriken Stoess, Germany), polyvinylalcohol (Mowiol® type 8–88, PVA, Hoechst, Germany), or carboxymethylcellulose sodium salt (Fluka, Germany). The *pH* of the gelatin containing solutions was adjusted to the isoelectric point (IEP) by adding dilute aqueous KOH. The specific charge and the IEP of the gelatin solutions were determined at different *pH* values by means of charge-compensating polyelectrolyte titration using a particle charge detector (PCD02 Mütek, Germany). The polyelectrolyte titration is based on the stoichiometric reaction between a polycation and a polyanion.

A typical carbon black dispersion is made up from *i)* 1% (w/w) Printex L6 (BET surface $265\,m^2 \cdot g^{-1}$, average primary particle size 18 nm), 2.7 m*M* sodium dioctylsulfosuccinate (AOT, Cyanamide), and 40 m*M* NaOAc or *ii)* 1% (w/w) Printex XE2 (BET surface $950\,m^2 \cdot g^{-1}$, average primary particle size 35 nm), 18 m*M* cetyltrimethylammonium bromide (*CTAB*, Aldrich), and 40 m*M* NaOAc, respectively. The preparation procedure of carbon black dispersions has been described in more detail previously [17, 18].

If not stated otherwise, PE and PP powders (typical particle size 300 μm) were conditioned for 2 min, rinsed with water for 2 min in a suction filter, stirred in the carbon black solution for 2 min, rinsed again with water for up to 1 min, and finally air-dried (Fig. 2).

The conductive composites were prepared by compression molding at the temperatures indicated in Figs. 6 and 7 and Tables 1 and 2. The resulting samples (diameter 6 cm) were flat and of homogeneous thickness (typically 1.5 mm).

The specific resistance of this kind of samples can be measured by the standard *van der Pauw* method which is applicable to flat samples of arbitrary shape if the contacts are sufficiently small and located at the circumferences of the sample [23].

In the present study, the amount of surfactant adsorbed on carbon black was determined quantitatively by polyelectrolyte titration and X-ray powder diffraction analysis. The spatial arrangement of molecular groups and ions is deduced from *Fourier* transform IR spectroscopy (FTIR).

X-Ray powder diffraction patterns of dried samples were measured with a Philips PW 1710 X-ray powder diffractometer with *Bragg-Brentano* geometry (CuK_{α_1} radiation, $\lambda = 1.5418$ Å). The spectra of the samples where a diffraction peak of crystalline *CTAB* was observed were calculated with the analysis software XPAS.

Samples of *CTAB* crystals and air-dried dispersions of carbon black were prepared as nujol or hexachlorobutadiene mulls between CaF_2 or CdTe windows. FTIR spectra of pure CTAB and dried dispersions were recorded with a *Fourier* transform PE-2000 spectrometer with a nominal resolution of $4\,cm^{-1}$.

Acknowledgements

This work was supported by the Austrian *Fonds zur Förderung der wissenschaftlichen Forschung* in the special research area *Elektroaktive Stoffe*.

References

[1] Fushimi K, Tsunakaw H, Yonahara K (1985) US Pat 4551267
[2] Tomacic G (1986) US Pat 4615108
[3] Kaun (1983) US Pat 4386019
[4] Herscovici C, Leo A, Charkey A (1988) US Pat 4758473
[5] Herscovici C (1990) US Pat 4920017
[6] Iemmi (1981) US Patent 4294893
[7] Balko (1982) US Patent 4339322
[8] Kordesch K, Gsellmann J, Jahangir S, Schautz M (1984) Proc Symp Porous Electrode: Theory and Practise, Detroit, MI, vol 84-8. Electrochemical Society, Pennington, NJ, pp 163–190
[9] Döring H, Clasen H, Zweynert M, Garche J, Jörissen L (1996) Materials for Bipolar Lead-Acid Batteries. In: Barsukov V, Beck F (eds) New Promising Electrochemical Systems for Rechargeable Batteries. NATO ASI Series, Kluwer, Dordrecht, pp 3–13
[10] Beck F (1996) Design and Materials for Metal-Free Rechargeable Batteries. In: Barsukov V, Beck F (eds) New Promising Electrochemical Systems for Rechargeable Batteries. NATO ASI Series, Kluwer, Dordrecht, pp 393–417
[11] Kühner K, Was ist Ruß? Degussa AG, Geschäftsbereich Anorganische Chemieprodukte, Frankfurt M, Germany
[12] Besenhard JO, Kühlkamp M, DT-PS 41 30 135
[13] Besenhard JO, Kühlkamp M, Huslage J (1996) Polypropylen/Rußverbundwerkstoffe als Mikroelektrodenessembles, GDCh-Monographie, Band 3, Gesellschaft Deutscher Chemiker, Frankfurt, p 461
[14] Mallory GO, Hajdu JB (1990) Electroless Plating, American Electroplaters and Surface Finishers Society
[15] Meyer H, Besenhard JO, Pejovnik S, Bele M, GB 9813733.4
[16] Besenhard JO, Claussen O, Gausmann H-P, Meyer H, Mahlkow H, US Pat 5 705 219
[17] Besenhard JO, Hanna S, Hagg Ch, Fiedler DA, Bele M, Pejovnik S, Meyer H (1997) Substrate induced coagulation: a method for coating surfaces with fine particulate materials. The Electrochemical Society Proceedings Series, 97-31, Interconnect and Contact Metallization, Rathore HS, Mathad GS, Plougonven C, Schuckert CC (ed) 96–107
[18] Bele M, Pejovnik S, Besenhard JO, Ribitsch V (1998) Substrate induced coagulation of carbon black on gelatin-modified printed wiring board surfaces, Part 1, Gelatin adsorption onto printed wiring board surfaces. Colloid Surface A **143**: 17
[19] Bele M, Kocevar K, Pejovnik S, Besenhard JO, Musevic I, Langmuir (submitted)
[20] Kocevar K, Bele M, Pejovnik S, Besenhard JO, Musevic I, Chem Phys Lett (submitted)
[21] Bele M, Kocevar K, Musevic I, Pejovnik S, Besenhard JO (2000) Colloid Surface A **168**: 231
[22] Sedin DL, Rowlen KL (2000) Anal Chem **72**: 2183
[23] van der Pauw LJ (1958) Philips Res Rep 13
[24] Besenhard JO, Hagg Ch, Hanna S, Meyer H, Deutsches Patent, März 1997 eingereicht, Aktenzeichen 197 12 103.9-45
[25] Haddadi-Asl V, Kazacos M, Skyllas-Kazacos M (1995) J Appl Electrochem **25**: 29
[26] Kim BS, Hayes A, Ralston J (1995) Carbon **33**: 25
[27] Hanna S (1999) Optimization of a Process for Coating Surfaces with Conductive Particles. PhD Thesis, Technische Universität Graz, Austria
[28] Bele M, Kodre A, Arcon I, Grdadolnik J, Pejovnik S, Besenhard JO (1998) Carbon **36**: 1207
[29] Abram JC, Bennett MC, (1968) J Colloid Interface Sci **27**: 1
[30] Medalia AI, Rivin D (1976) In: Parfitt GD, Sing KSW (eds) Characterisation of Powder Surfaces, chapt 7. Academic Press, London, pp 334–335
[31] Hess WM, Herd CR (1993) In: Donnet JB, Bansal RC, Wang WJ (eds) Carbon Black Science and Technology, chapt 3. Dekker, New York, pp 118–119

[32] Celik MS (1998) J Colloid Interface Sci **129**: 428
[33] Sasaki Y, Miyassu YI, Lee S, Nagadome S, Igimi H, Sugihara G (1996) Colloid Surface B-Biointerfaces **7**: 181
[34] Besenhard JO, Meyer H, Gausmann HP, Mahlkow H, PCT/DE 92/00315
[35] Besenhard JO, Meyer H, Gausmann HP, Ger Pat 41 13 407
[36] Besenhard JO, Claußen O, Gausmann HP, Meyer H, Ger Pat 41 41 416
[37] DIN 53 455-Vorschrift
[38] DIN 53 457-Vorschrift

Received August 24, 2000. Accepted September 19, 2000

Application of the *van der Pauw* Method to Conductivity Relaxation Experiments on $YBa_2Cu_3O_{6+\delta}$

Wolfgang Preis[1], **Michael Holzinger**[2], and **Werner Sitte**[2,*]

[1] Institut für Physikalische Chemie, Montanuniversität Leoben, A-8700 Leoben, Austria
[2] Institut für Physikalische und Theoretische Chemie, Technische Universität Graz, A-8010 Graz, Austria

Summary. The *van der Pauw* method has been applied to conductivity relaxation experiments on $YBa_2Cu_3O_{6+\delta}$ at 600°C in order to determine the chemical diffusion coefficient as a function of the oxygen partial pressure in the surrounding atmosphere $(10^0 > p_{O_2}/\text{bar} > 10^{-3})$. It is shown that the *van der Pauw* technique is suitable for monitoring the conductivity relaxation when the oxygen diffusion is perpendicular to the direct current flowing through the sample in accordance with the *van der Pauw* geometry using thin tablets as samples. The oxygen partial pressure is changed stepwise (generally $\Delta \log p_{O_2} \leq 0.5$) by employing appropriate gas mixtures as well as an electrochemical oxygen pump device. An evaluation formula is given for the determination of the chemical diffusion coefficient neglecting slow surface processes. In addition, the electronic conductivity of $YBa_2Cu_3O_{6+\delta}$ has been measured at 600°C as a function of oxygen partial pressure of the ambient atmosphere $(10^0 > p_{O_2}/\text{bar} > 10^{-5})$ by means of the *van der Pauw* method applying the same experimental set-up. Typical values of the chemical diffusion coefficient are in the range of $10^{-6}\,\text{cm}^2\cdot\text{s}^{-1}$; the results of the conductivity measurements are interpreted in terms of an appropriate defect model.

Keywords: *van der Pauw* method; Conductivity relaxation; Chemical diffusion; $YBa_2Cu_3O_{6+\delta}$.

Introduction

Complex investigations of oxides require the availability of various experimental techniques. For the defect chemical modeling of the oxides usually electronic conductivity measurements as a function of the oxygen activity are employed [1, 2]. Conductivity relaxation experiments yield the chemical diffusion coefficient and – if necessary – the oxygen exchange coefficient in contrast to *e.g.* $^{18}O/^{16}O$-tracer-SIMS experiments which give the self-diffusion coefficient of oxygen [3].

Conductivity relaxation is an adequate method to determine chemical diffusion coefficients of oxides. As electrical conductivity is much more sensitive to changes

* Corresponding author

of the oxygen chemical potential in the atmosphere than *e.g.* weight, considerable changes in conductivity can be observed even for small p_{O_2} steps. On the one hand this makes the electrical conductivity relaxation technique more easily applicable to a wide variety of materials; on the other hand, conductivity relaxation can only be applied to oxides when defect chemistry allows to assume a correlation between the change of the oxygen content of the specimen and the change in electrical conductivity. Usually, this correlation is a linear one, but this is not a necessity as even in a mixed p/n regime, where linear correlation fails, conductivity relaxation can be used [4, 5]. Additionally, it has to be mentioned that the transport properties of mixed oxides may become complex when internal defect (redox) reactions have to be taken into account [6].

In recent years, conductivity relaxation has been applied to a number of perovskites for use as high temperature electrochemical devices (oxygen electrodes or separation membranes) [7–11]. Usually, bar-shaped samples have been employed to determine the chemical diffusion coefficient and the oxygen surface exchange coefficient [12]. Alternatively, in this study we apply the *van der Pauw* technique [13] for conductivity relaxation measurements. The *van der Pauw* technique permits measurements of the electronic conductivity of mixed ionic-electronic conductors even if the ionic conductivity is of the same order as the electronic one. Nevertheless, with electronic conductivity measurements on mixed conductors the activity gradient created during the conductivity measurements has to be taken into account in order to increase the stoichiometric resolution of the results [14]. The oxygen partial pressure of the gaseous atmosphere surrounding the sample can be controlled over a wide range (10^{-26} bar $\leq p_{O_2} \leq 1$ bar) using either gas mixtues (*e.g.* Ar/O_2, H_2/H_2O, or CO/CO_2 mixtures obtained by flow controllers) and/or an electrochemical oxygen pump device (based on a zirconia cell).

It is the aim of this paper to apply the *van der Pauw* geometry for conductivity relaxation measurements on mixed conducting oxides, especially $YBa_2Cu_3O_{6+\delta}$. For these kinds of measurements we employ the same experimental set-up as for electronic conductivity measurements. The oxygen partial pressure of the carrier gas surrounding the sample is changed stepwise. The oxygen content of the sample attains a new equilibrium state depending on surface exchange and chemical diffusion coefficients. The conductivity relaxation is measured perpendicularly to the direction of oxygen diffusion in the sample according to the *van der Pauw* geometry, and the conductivity data are fitted to the solution of *Fick*'s second law under the relevant initial and boundary conditions. It must be mentioned that for all our measurements we assumed that the oxygen surface exchange coefficient was by orders of magnitude larger than the chemical diffusion coefficient. It is well known that the oxygen exchange kinetics of $YBa_2Cu_3O_{6+\delta}$ is very fast. This may be concluded *e.g.* from the fact that $YBa_2Cu_3O_{6+\delta}$ electrodes allow the oxygen exchange from $SrTiO_3$ samples even at very low temperatures (423 K) so that not only the electrode semicircle disappears (as indicated by impedance measurements), but also the correct p_{O_2} dependence of the bulk conductivity of $SrTiO_3$ can be found [15]. Additionally, we had good evidence for the absence of surface exchange as samples of different thickness gave identical values of the chemical diffusion coefficient.

Theory

When the oxygen partial pressure in the ambient atmosphere of a mixed conducting oxide ceramic is changed stepwise, the oxygen content of the mixed conductor is adjusted in order to attain a new thermodynamic equilibrium between the ambient atmosphere and the ceramic. This relaxation process is accompanied with oxygen exchange at the surface and chemical diffusion in the bulk of the mixed conducting oxide. The chemical diffusion process is described by *Fick*'s second law (Eq. (1)).

$$\frac{\partial c}{\partial t} = \tilde{D}\frac{\partial^2 c}{\partial x^2} \tag{1}$$

The chemical diffusion coefficient $\tilde{D}$ is assumed to be spatially invariant. The symbol c denotes the concentration of the diffusing species which corresponds to the stoichiometric excess or deficiency of oxygen in the oxide ceramic. The rate of oxygen exchange at the surface of the oxide is given by the difference between the chemical potential of oxygen μ_O in the surrounding atmosphere and the oxide ceramic. If the oxygen partial pressure step is sufficiently small, a linear approximation of the exchange rate will be fulfilled. Thus, assuming the ceramic sample to be a thin slab with parallel surface planes, the boundary conditions for the diffusion equation (1) read [10]

$$-\tilde{D}\frac{\partial c}{\partial x} = \tilde{k}(c_\infty - c);\ x = 0 \tag{2a}$$

and

$$\frac{\partial c}{\partial x} = 0;\ x = \frac{L}{2} \tag{2b}$$

with $\tilde{k}$, c_∞ and L representing the exchange coefficient, the equilibrium surface concentration of oxygen at the end of the relaxation process, and the length of the sample, respectively. If the surface exchange coefficient of the ceramic is higher by orders of magnitude than the chemical diffusion coefficient ($\tilde{k} \gg \tilde{D}$), the surface of the specimen can be assumed to be in equilibrium with the ambient atmosphere during the whole relaxation experiment. Therefore, the boundary conditions can be simplified to

$$c = c_\infty;\ x = 0 \tag{3a}$$

and

$$\frac{\partial c}{\partial x} = 0;\ x = \frac{L}{2}. \tag{3b}$$

Applying a standard technique (*e.g. Laplace* transformation to Eqs. (1), (3a), and (3b), the solution function can be written as shown in Eq. (4) where c_O denotes the initial diffusant concentration at $t = 0$ [16].

$$c(x) = c_O + (c_\infty - c_O) \times \left(1 - \frac{4}{\pi}\sum_{n=0}^{\infty}\frac{\cos((2n+1)\pi(x/L - 1/2))\exp(-(2n+1)^2\pi^2\tilde{D}t/L^2)}{(-1)^n(2n+1)}\right) \tag{4}$$

The electrical conductivity change of an oxide ceramic owing to oxygen absorption or desorption is related to the variation of the oxygen content of the sample by the charge neutrality condition. When the defect chemistry of the ceramic allows one to assume an almost linear relationship between the change of the oxygen content of the specimen and the charge carrier concentration, the change of the conductivity $\Delta\sigma$ of the specimen is likewise proportional to the variation of the diffusant concentration with time ($\Delta\sigma \propto \Delta c$). Hence, the chemical diffusion coefficient of such compounds can be easily determined by monitoring the relaxation of the electronic conductivity of the sample.

The *van der Pauw* method is a wide-spread four point technique for conductivity measurements [13, 14, 17, 18]. For the determination of the electronic conductivity, four ionically blocking contacts (usually platinum) are placed peripherally on the sample (Fig. 1). A constant direct current I_{RS} is fed through the electrodes R and S,and the voltage U_{PQ} is measured between the probes P and Q. The sample can be assumed to consist of parallel layers of infinitesimal thickness dx showing a uniform conductivity $\sigma(x)$. As both the probe voltage U_{PQ} and the position of the electrodes and probes are equal for all parallel layers, the relation between the probe voltage and the current dI_{RS} for each of these layers is given by

$$\mathrm{d}I_{RS} = k^{-1} U_{PQ} \sigma(x) \mathrm{d}x; \; k = \frac{1}{\pi} \ln\left(\frac{(z_S - z_Q)(z_R - z_P)}{(z_R - z_Q)(z_S - z_P)}\right) \tag{5}$$

z_R, z_S, z_P, and z_Q correspond to the positions of the electrodes and probes in the semi-infinite plane which is obtained by application of conformal mapping to the sample geometry [14]. If the specimen is a thin tablet (circular geometry, Fig. 1), the pertinent coordinates are given by

$$z = \mathrm{i}\frac{1 + \mathrm{e}^{\mathrm{i}\varphi}}{1 - \mathrm{e}^{\mathrm{i}\varphi}} = -\frac{\cos(\varphi/2)}{\sin(\varphi/2)}. \tag{6}$$

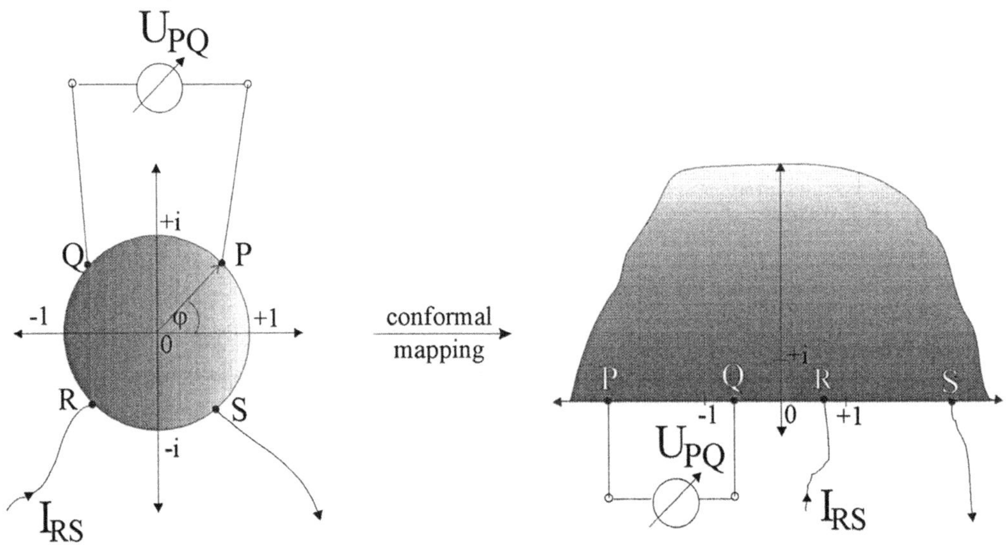

Fig. 1. Schematic drawing of the *van der Pauw* configuration; conformal mapping allows the circular sample to be transformed into a semi-infinite plane

When equidistant electrodes and probes are employed ($\varphi_P = -\varphi_S = \pi/4$; $\varphi_Q = -\varphi_R = 3\pi/4$), the relevant positions amount to $z_S = -z_P = 2.4142$ and $z_R = -z_Q = 0.4142$, and the geometric factor k^{-1} in Eq. (5) is equal to $\pi/\ln 2$. The total current flowing through the mixed conducting ceramic is given by

$$I_{RS} = k^{-1} U_{PQ} \int_{x=0}^{L} \sigma(x) \mathrm{d}x. \tag{7}$$

If the oxide ceramic shows a linear relationship between $\Delta\sigma$ and the change of oxygen excess or deficiency with time, Eq. (7) can be written as

$$I_{RS} = k^{-1} U(t) \int_{x=0}^{L} \left(\frac{c(x) - c_0}{c_\infty - c_0} (\sigma_\infty - \sigma_0) + \sigma_0 \right) \mathrm{d}x \tag{8}$$

with σ_∞ and σ_0 being the conductivity of the sample at the end of the relaxation process and at the initial time $t = 0$, respectively. For the sake of simplicity we omit the index "$_{PQ}$" in the symbol U_{PQ} from now on. Inserting Eq. (4) into Eq. (8) and evaluating the integral, we obtain

$$I_{RS} = k^{-1} U(t) L \left(\left(1 - \frac{8}{\pi^2} \sum_{n=0}^{\infty} \frac{\mathrm{e}^{-(2n+1)^2 \pi^2 \tilde{D} t / L^2}}{(2n+1)^2} \right) (\sigma_\infty - \sigma_0) + \sigma_0 \right). \tag{9}$$

In the case of $t = 0$, Eq. (9) reads

$$I_{RS} = k^{-1} U(0) L \sigma_0, \tag{10a}$$

and for $t \to \infty$ we find

$$I_{RS} = k^{-1} U(\infty) L \sigma_\infty. \tag{10b}$$

After some basic algebra, a combination of Eqs. (9), (10a), and (10b) leads to

$$\frac{(U(t) - U(\infty)) U(0)}{(U(0) - U(\infty)) U(t)} = \frac{8}{\pi^2} \sum_{n=0}^{\infty} \frac{\exp(-(2n+1)^2 \pi^2 \tilde{D} t / L^2)}{(2n+1)^2} \tag{11}$$

where $U(0)$ and $U(\infty)$ denote the probe voltage at the begin and the end of the relaxation experiment, respectively. The direct current I_{RS} is kept constant during the relaxation experiment. When the relaxation times are fairly long, only the first term in the series of Eq. (11) has to be taken into account, yielding

$$\ln\left(\frac{U(t) - U(\infty)}{U(0) - U(\infty)} \right) - \ln\left(\frac{U(t)}{U(0)} \right) = \ln\left(\frac{8}{\pi^2} \right) - \frac{\tilde{D}\pi^2}{L^2} t. \tag{12}$$

A plot of the left-hand side of Eq. (12) *vs.* time should result in a straight line at fairly long relaxation times; the chemical diffusion coefficient is obtained from its slope as $\tilde{D}\pi^2/L^2$.

Results and Discussion

In Fig. 2 the electronic conductivity of $YBa_2Cu_3O_{6+\delta}$ is plotted against the oxygen partial pressure at 600°C. The results are in good agreement with a defect model

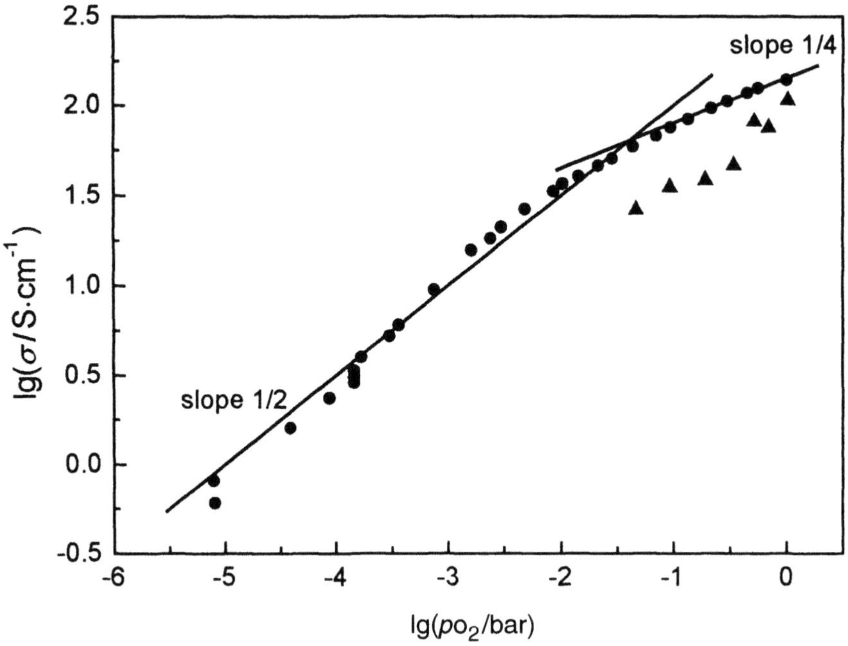

Fig. 2. Electronic conductivity of $YBa_2Cu_3O_{6+\delta}$ as a function of the oxygen pressure at 600°C; the results of *Maier et al.* [20] are included (solid triangles)

proposed by *Maier et al.* [20] where all excess oxygen atoms are interstitial defects O_i'. In a medium oxygen partial pressure range ($10^{-5} < p_{O_2}/\text{bar} < 10^{-2}$) the concentration of O_i' is assumed to be independent of the oxygen partial pressure of the surrounding atmosphere. Thus, the neutrality condition reads $[Y_{Ba}^\bullet] = [O_i']$ with the concentration of holes (defect electrons) $h^\bullet$ being smaller by orders of magnitude than the concentration of the defects $Y_{Ba}^\bullet$. According to the equilibrium

$$V_i^x + 1/2O_2 \rightleftharpoons O_i' + h^\bullet \rightleftharpoons O_i^x, \tag{13}$$

the hole concentration varies linearly with the concentration of O_i^x. As the oxygen excess δ is given by the sum of the mole fractions of O_i' and O_i^x, a linear relationship between the electronic conductivity σ and the oxygen content is expected:

$$c = \frac{\delta}{V_m} = [Y_{Ba}^\bullet](1 + K[h^\bullet]). \tag{14}$$

K and V_m denote the equilibrium constant of the defect reaction $O_i' + h^\bullet \rightleftharpoons O_i^x$ and the molar volume of the mixed conducting ceramic, respectively. Combining Eq. (14) with the relation $\sigma = Fb_h[h^\bullet]$ results in

$$\Delta\sigma = \frac{Fb_h}{[Y_{Ba}^\bullet]K}\Delta c = \text{const} \cdot \Delta c \tag{15}$$

where F and b_h denote the *Faraday* constant and the mobility of the holes, respectively. Hence, Eq. (12) is valid for the determination of the chemical diffusion coefficient of $YBa_2Cu_3O_{6+\delta}$ in a medium oxygen partial pressure range. It is worth

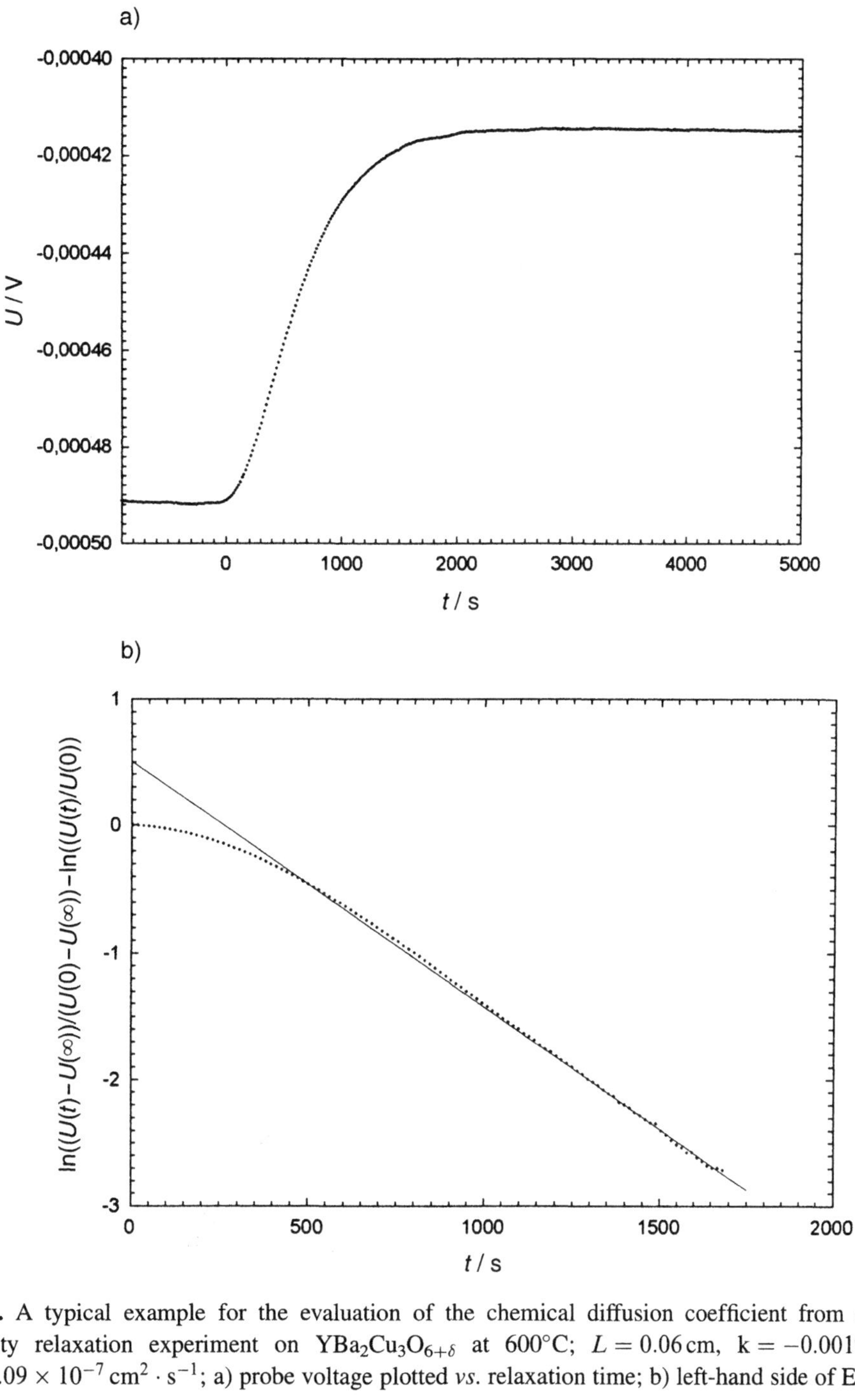

Fig. 3. A typical example for the evaluation of the chemical diffusion coefficient from a conductivity relaxation experiment on $YBa_2Cu_3O_{6+\delta}$ at 600°C; $L = 0.06\,\text{cm}$, $k = -0.00194\,\text{s}^{-1}$, $\tilde{D} = 7.09 \times 10^{-7}\,\text{cm}^2 \cdot \text{s}^{-1}$; a) probe voltage plotted *vs.* relaxation time; b) left-hand side of Eq. (12) plotted *vs.* relaxation time; the chemical diffusion coefficient is obtained from the slope of the linear part of the graph

mentioning that in accordance with Eq. (13) the hole concentration as well as the electronic conductivity vary with the square root of oxygen partial pressure ($\sigma \propto [h^\bullet] \propto p_{O_2}^{1/2}$) as can be seen in Fig. 2. At high p_{O_2} values ($p_{O_2} > 0.1\,\text{atm}$) the concentration of the holes increases, leading to a new neutrality condition

($[h^\bullet] = [O_i']$), and the variation of the conductivity with the oxygen partial pressure shifts to $\sigma \propto p_{O_2}^{1/4}$ (see Fig. 2). In this case, the relation between the oxygen content and the hole concentration in $YBa_2Cu_3O_{6+\delta}$ is given by

$$c(x) = \frac{\delta}{V_m} = [h^\bullet] + K \cdot [h^\bullet]^2. \tag{16}$$

Again, combining Eq. (16) with the relation $\sigma = Fb_h[h^\bullet]$ yields

$$\Delta\sigma = \frac{Fb_h}{1 + 2\sigma_0 K F^{-1} b_h^{-1}} \Delta c = \text{const} \cdot \Delta c; \;\; \Delta\sigma \ll \sigma_0 \tag{17}$$

with σ_0 being the conductivity at the initial time $t = 0$. Therefore, the change of the conductivity ($\Delta\sigma$) with time can be assumed to vary linearly with Δc when the oxygen partial pressure step is kept sufficiently small.

A typical example for a conductivity relaxation experiment at 600°C is illustrated in Fig. 3. The probe voltage is plotted as a function of time in Fig. 3a. The increase of the voltage is caused by an instantaneous change of the oxygen partial pressure in the ambient atmosphere from 2.7 to 4.0 mbar. A plot of the left-hand side of Eq. (12) *vs.* time is shown in Fig. 3b. A straight line occurs at fairly long relaxation times, allowing the evaluation of the chemical diffusion coefficient from the slope. The variation of the chemical diffusion coefficient of $YBa_2Cu_3O_{6+\delta}$ with the oxygen partial pressure at 600°C is shown in Fig. 4. The diffusion coefficient increases with increasing oxygen content of the sample with typical values around $10^{-6}\,cm^2 \cdot s^{-1}$. Furthermore, it should be mentioned that our results

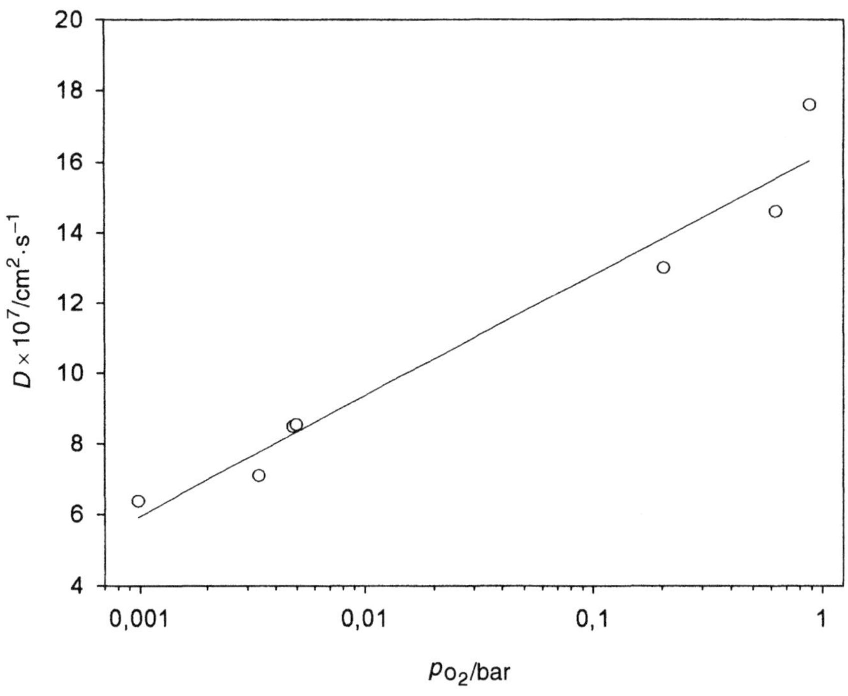

Fig. 4. Chemical diffusion coefficient $\tilde{D}$ of $YBa_2Cu_3O_{6+\delta}$ as a function of oxygen pressure at 600°C

are in good agreement with the data of *Park et al.* [21] as well as *LaGraff* and *Payne* [22].

Conclusions

A conductivity relaxation method was applied to $YBa_2Cu_3O_{6+\delta}$ at 600°C in order to determine the chemical diffusion coefficient as a function of oxygen partial pressure of the ambient atmosphere. The oxygen partial pressure was changed stepwise, and chemical diffusion coefficients were determined in the range of $10^0 - 10^{-3}$ bar. It was shown that the *van der Pauw* technique is suitable for monitoring the conductivity relaxation when the oxygen diffusion is perpendicular to the direct current flowing through the sample according to the *van der Pauw* geometry. Obviously, this restriction is fulfilled when the specimen is a thin tablet. A linear relation (Eq. (12)) was derived in order to determine the chemical diffusion coefficient from the experimental data disregarding slow surface processes, since the surface exchange coefficient is higher by orders of magnitude than the chemical diffusion coefficient in the case of $YBa_2Cu_3O_{6+\delta}$. In addition, the electronic conductivity of $YBa_2Cu_3O_{6+\delta}$ was measured as a function of oxygen partial pressure ($10^{-5} < p_{O_2}/\text{bar} < 10^0$), and the results were interpreted in terms of a defect model published recently [20]. It should be emphasized that the *van der Pauw* method was used for the determination of both the chemical diffusion coefficient and the electronic conductivity as a function of oxygen partial pressure of the surrounding atmosphere. Thus, the same experimental set-up allows the measurement of both quantities which are of considerable importance for a better understanding of the transport properties of oxide ceramics.

Experimental

$YBa_2Cu_3O_{6+\delta}$ sample was prepared from $YBa_2Cu_3O_{6+\delta}$ powder (Davison Chemical, Chattanooga, Tennessee, USA) which was pressed to a disk of 10 mm diameter and 0.60 mm thickness and annealed at 900°C in air for 2 days. The sample was then placed in a sealed quartz apparatus in a high temperature furnace. Four platinum contacts were positioned on one face of the sample with the help of light spring action. The oxygen activity of the carrier gas surrounding the sample was varied over a wide oxygen partial pressure range (10^{-6} bar $\leq p_{O_2} \leq 1$ bar) using either appropriate gas mixtures (argon 99.999% and oxygen 99.95%, admixed by flow controllers) or an electrochemical oxygen pump device consisting of an yttria stabilized zirconia tube (Zircoa, USA) with porous platinum contacts on the inner and outer side. This device allowed to pump oxygen into or out of the flowing gas under galvanostatic conditions (operated at 700°C). A zirconia microgauge [19] was used to measure the oxygen activity of the carrier gas. p_{O_2} changes were kept as small as possible (generally $\Delta \log p_{O_2} \leq 0.5$). Direct currents were provided by precision current sources (Knick J152), and voltages were recorded using nanovoltmeters (Keithley 182) and high-impedance multimeters (Keithley 199). Temperatures were kept constant using a precision temperature controller (Eurotherm 818; Pt-Pt10%Rh-thermocouples). All data were recorded by a computer controlled data acquisition system.

Acknowledgements

We express our thanks to the *Austrian Science Foundation* (FWF) for supporting this work within the Special Research Program 'Electroactive Materials' (Project F0915).

References

[1] Kofstad P (1972) Nonstoichiometry, Diffusion and Electrical Conductivity in Binary Oxides. Wiley-Interscience
[2] Sørensen OT (ed) (1981) Nonstoichiometric Oxides. Academic Press
[3] Chater RJ, Carter S, Kilner JA, Steele BCH (1992) Solid State Ionics **53–55**: 859
[4] Song CR, Yoo HI (1999) Solid State Ionics **120**: 141
[5] Song CR, Yoo HI (1999) Solid State Ionics **124**: 289
[6] a) Maier J (1993) J Am Ceram Soc **76**: 1212; b) Maier J (1993) J Am Ceram Soc **76**: 1218; c) Maier J (1993) J Am Ceram Soc **76**: 1223; d) Maier J (1993) J Am Ceram Soc **76**: 1228
[7] Yasuada I, Hishinuma M (1995) Solid State Ionics **80**: 141
[8] Yasauda I, Hishinuma M (1995) J Solid State Chem **115**: 152
[9] ten Elshof JE, Lankhorst MHR, Bouwmeester HJM (1997) Solid State Ionics **99**: 15
[10] ten Elshof JE, Lankhorst MHR, Bouwmeester HJM, Verweij H (1997) J Electrochem Soc **144**: 1060
[11] Jacobson AJ, Kim S, Wang S, Yang YL, Abeles B (1999) Proc 12th Intern Conf on Solid State Ionics. Halkidiki, Greece, p 240
[12] Maier J (1998) Solid State Ionics **112**: 197
[13] van der Pauw LJ (1958) Philips Res Rep **13**: 1
[14] Grientschnig D, Sitte W (1990) Z Phys Chem N F **168**: 143
[15] Denk I, Claus J, Maier J (1997) J Electrochem Soc **144**: 3526
[16] Carslaw HS, Jaeger JC (1959) Conduction of Heat in Solids. Clarendon Press, Oxford
[17] Riess I, Tannhauser DS (1982) Solid State Ionics **7**: 307
[18] Preis W, Sitte W (1995) Solid State Ionics **76**: 5
[19] Fouletier J, Vitter G, Kleitz M (1975) J Appl Electrochem **5**: 111
[20] Maier J, Murugaraj P, Pfundtner G, Sitte W (1989) Ber Bunsenges Phys Chem **93**: 1350
[21] Park JH, Kostic P, Singh JP (1988) Mater Letts **6**: 393
[22] LaGraff J, Payne DA (1993) Phys Rev B **47**: 3380

Received May 30, 2000. Accepted June 8, 2000

Thin Films and Interfaces of Electroactive Organic Materials: The Surface Science Approach

Robert I. R. Blyth, Renate Duschek, Georg Koller, Falko P. Netzer, and Michael G. Ramsey*

Institut für Experimentalphysik, Karl-Franzens-Universität Graz, A-8010 Graz, Austria

Summary. We highlight the importance of interfacial properties in determining the performance of devices based on electroactive organic materials. Investigations of the interfaces of benzene with Al(111) and In_2O_3 are presented as a model of interface properties for devices based on complex aromatic molecules. At both interfaces the binding is shown to be electrostatic, with the resulting interface dipole determining the band alignment. It is also argued that chemical modification of substrates can be used to tailor both electronic and structural properties.

Keywords. Electronic structure; Electroactive materials; Photoemission spectroscopy; Surface.

Introduction

The nature of the interface of large organic molecules with inorganic materials is of growing technological importance. In particular, for π-conjugated molecules the pace of development in device applications, *e.g.* field effect transistors (FETs), light emitting diodes (LEDs), and even lasers, is far outstripping the advances in the understanding of the critically important electrode interfaces. Whereas the organic/inorganic interface was recognized early as an important factor in governing the macroscopic properties of devices, there has been little advance in understanding why.

Over the last thirty years a wealth of ultra-high vacuum (UHV) techniques have been developed to examine surfaces and interfaces [1]. The development of these so-called surface science spectroscopies was to a great extent driven by three industries: materials science which needed to characterize surfaces, heterogeneous catalysis which needed an understanding of the nature of the chemical bond at the molecule/inorganic solid interface, and microelectronics which needed a window in the electronic properties of the conductor/inorganic semiconductor interface. The issues addressed by these endeavours – interfacial characterization, chemistry, and electronic structure – are indeed those required for a microscopic understanding of

* Corresponding author

the interfaces of devices based on active organic materials. Such an understanding is necessary to bring device performance to a reproducible optimum level.

These basic issues are best illustrated by considering a typical simple prototype light emitting diode (LED) consisting of a thin film of the oligomer sexiphenyl sandwiched between an electron injecting anode of aluminum and a transparent hole injecting electrode (tin-doped indium oxide known as indium-tin-oxide, *ITO*). Although produced in high vacuum, the macroscopic properties of the device are far from reproducible with, for instance, the onset voltage for blue light emission varying from a useful 4 V to an unacceptable 15 V depending on the rate at which the aluminum was deposited [2]. A chemical cross-section of such a device can be achieved by performing an *Auger* depth profile, whereby *Auger* electron spectra are recorded continuously while etching through the device by inert ion bombardment (sputtering).

A representative cross-section is shown in Fig. 1. The use of *ex vacuo* prepared substrates together with the relatively high background pressures during growth ($\sim 10^{-6}$ mbar of water) ensures that the interface composition is essentially uncontrolled. In particular, there is much hydrocarbon contamination of the (nanoscopically rough) *ITO* surface, and there is oxygen at the Al/sexiphenyl interface. By comparing cross sections of devices and their macroscopic properties, it is clear that the presence of oxygen at the Al/sexiphenyl interface improves the performance. It is therefore necessary to understand the interface properties at the microscopic level in order to comprehend device performance at the macroscopic level. This requires experiments under ultra-high vacuum (UHV), *i.e.* pressures below 10^{-9} mbar to ensure cleanliness and allow precise control of the interfaces. *In situ* experiments with model interfaces allow the use of numerous surface spectroscopies such as angle-resolved UV photoemission (ARUPS), X-ray photoemission (XPS), and thermal desorption (TDS). Single crystal surfaces of known geometric and electronic structure can be readily prepared atomically clean and provide suitable models for the electrodes. In this work, Al(111) has been used to model the ubiquitous Al electrode of organic devices. Commercial *ITO* is unsuited

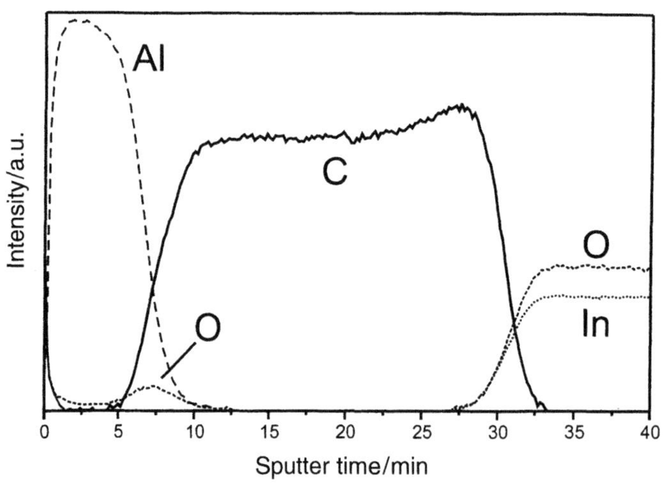

Fig. 1. *Auger* depth profile of a sexiphenyl based LED with aluminum and *ITO* electrodes

to controlled surface science experiments as it cannot be readily or reproducibly cleaned in UHV. A model substrate for the *ITO* electrode has therefore to be found. In_2O_3 itself is suitable in some respects, but is of course an insulator and would therefore cause charging problems during electron spectroscopy. However, it is possible to grow an ultra-thin film (thickness in the order of a few atomic layers) of stoichiometric In_2O_3 by controlled oxidation of a NiIn alloy [3]. This method of preparing ultra-thin oxide films, which relies on favourable thermodynamics to preferentially oxidize only one component of the alloy, was developed to model oxide supports for catalyst studies [4]. Benzene is the simplest model for aromatic-based (semiconducting) polymers. Thin films can be formed while following chemistry and electronic structure from submonolayer to multilayer coverages. In this work we have studied benzene on Al(111) and In_2O_3 to model the two interfaces of the device of Fig. 1.

If devices based on organic materials are ever to have a significant impact in the marketplace there will be a need to control interface formation and film growth on the nanoscopic scale, as is the case at present with inorganic semiconductor devices. Such molecular architecture, *via* control of the chemical and geometric structure of the substrate, and careful choice of the adsorbate functional groups will allow tailoring of the properties of the interface. Research in the area of ordered organic overlayers has largely been focused in the areas of self-assembly and *Langmuir-Blodgett* films. In both cases the order derives from the dominance of molecule-molecule interactions over molecule-substrate interactions. From the point of view of controlling the interfaces at the sub-nanometre scale, a potentially more rewarding route is *via* substrate templating: here the aim is to structure the substrate in such a way as to predetermine the bonding site and thereby impose order on the organic overlayer. This forced assembly has been recently demonstrated for bithiophene on sulfur-modified Ni(110) [5]; the structuring, however, was on too small a scale (stripes with 10 Å periodicity) for any near future technologies. Oxygen induced reconstructions do, however, occur on numerous metal surfaces on a more mesoscopic scale. In this contribution, we have investigated whether Cu(110)p(2×1)-O striped structures can act as templates for organic film growth.

Results and Discussion

We consider first the benzene/Al interface. Benzene does not adsorb at room temperature, and so dosing was performed with the Al substrate held at a temperature of 90 K. Thermal desorption spectra – monitoring the benzene signal in the mass spectrometer while ramping the sample temperature – from increasing doses of benzene on Al(111) are shown in Fig. 2. Note that the amount of benzene dosed can be precisely controlled; the units of exposure are *Langmuirs* (*L*), where $1\,L = 1 \times 10^{-6}\,\text{mbar} \cdot \text{s}$. Desorption from the multilayer, which is held together by weak *van der Waals* type interactions, occurs at 145 K. The benzene monolayer, *i.e.* the benzene in contact with the Al surface, desorbs only 6 K higher. The absence of a strong interaction might be considered surprising, given the reactive nature of the Al surface. By comparison, monolayers of benzene on transition metals are stable above room temperature [6, 7]. ARUPS spectra can give the band alignment directly (HOMO with respect to the *Fermi* level) and also allow the nature of the

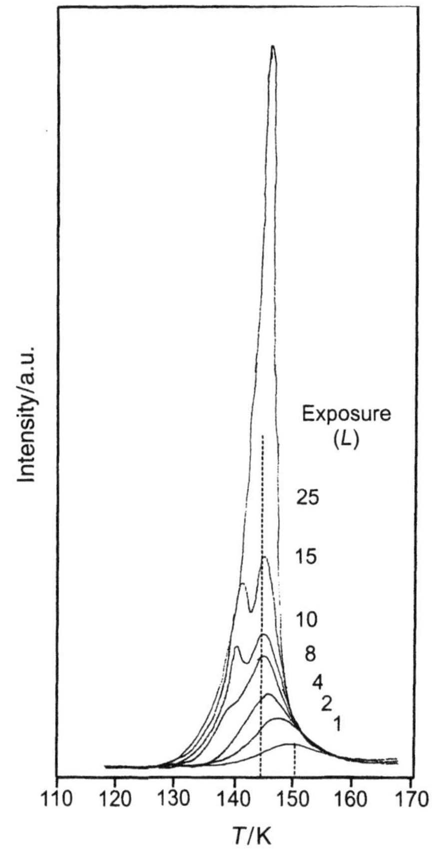

Fig. 2. Thermal desorption spectra of increasing doses of benzene on Al(111) at 90 K; exposures are in *Langmuirs* (*L*) where $1\,L = 1 \times 10^{-6}$ mbar · s

chemical bond to be assessed. In Fig. 3 the ARUPS spectra of a monolayer of benzene on Al(111) at two different emission angles are shown together with the spectrum of the subsequently grown multilayer for comparison. In the multilayer situation there is no long-range order and no strong intermolecular interaction, and so the spectrum is analogous to that of gas phase benzene [8], allowing identification of the various valence band orbitals. The correlation between the orbital emissions of monolayer and multilayer are also indicated in Fig. 3. The HOMO in benzene is the $1e_{1g}$-π-orbital, whose binding energy relative to the Al Fermi level can be readily quantified from Fig. 3. The secondary electron cut off the photoemission spectra (not shown) has been used to determine the work function change as benzene is adsorbed which is shown in the inset of Fig. 3. Note that there are no differential orbital energy shifts in the monolayer relative to the multilayer. This implies that the surface bond has not induced any significant changes to the molecular electronic structure. Specifically, there is no π-stabilization, and thus no π-bonding, in contrast to the general case on transition metals [6, 7, 9]. Figure 3 also shows that in the monolayer different peaks are suppressed at different emission angles, which enables the use of spectroscopic selection rules to determine its symmetry. A full analysis is beyond the scope of this paper, but it can be shown [10]

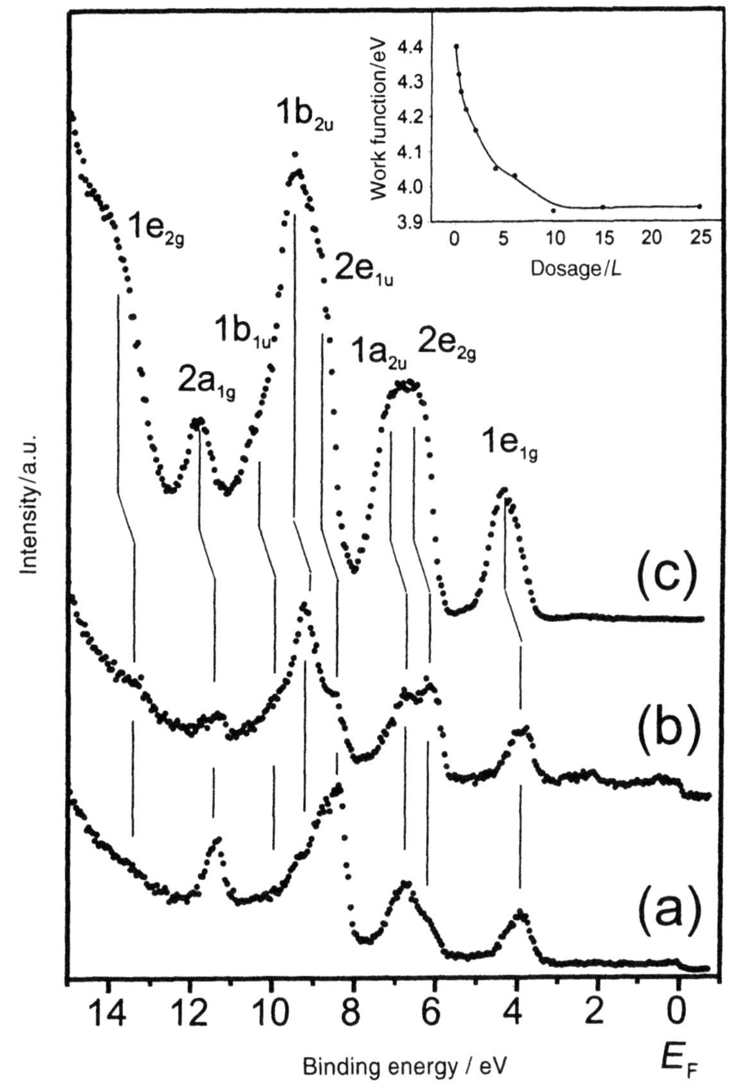

Fig. 3. ARUPS spectra (He I, $h\nu = 21.2\,\text{eV}$) of benzene on Al(111); (a) monolayer, normal emission; (b) monolayer, 60° emission; (c) multilayer; the inset shows the work function change as benzene is adsorbed

that in this case the molecular symmetry is C_{6V}, which implies that the molecule lies flat on the surface and is barely perturbed since it does not adopt the C_{3V} substrate symmetry. Recent local density functional theory calculations [10] indicate that the bond is electrostatic with a dipole induced in the Al(111) surface. It is this dipole which determines the work function change and governs the band alignment.

Turning now to the other interface, ARUPS spectra of benzene on the In_2O_3 film are shown in Fig. 4. The identification of the HOMO is more difficult than was the case on Al(111) due to the presence of substrate features in the spectrum, as can be seen in Fig. 4a which shows the spectrum of the clean In_2O_3 film. The In_2O_3 valence band is known from XPS measurements [11] to be 6 eV wide, but only the

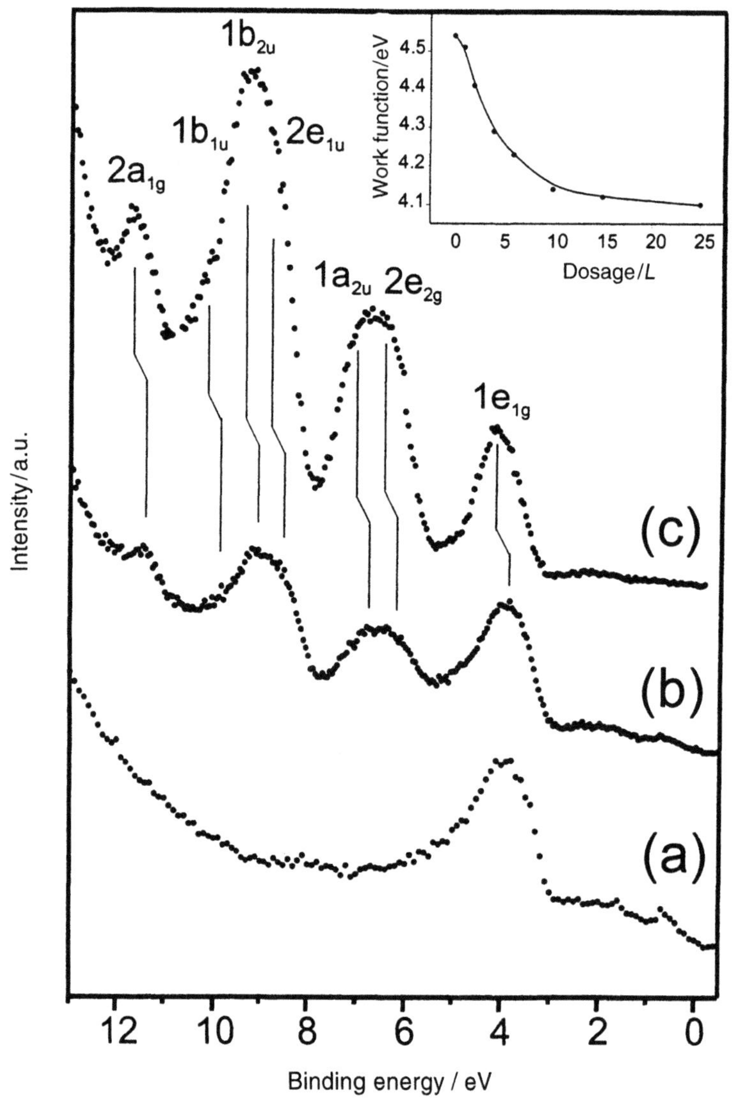

Fig. 4. Normal emission ARUPS spectra (He I, $h\nu = 21.2\,\text{eV}$) of (a) clean In_2O_3/NiIn, (b) with a monolayer of benzene, and (c) with a multilayer of benzene; the inset shows the work function change as benzene is adsorbed

leading edge is visible using He I radiation, giving rise to a peak at about 4 eV. However, the absence of any asymmetry in the monolayer peak at 4 eV, together with the absence of a differential shift in the other π-orbital ($1a_{2u}$) implies that the $1e_{1g}$ energy is *ca.* 3.9 eV in the monolayer. Unlike the case on Al(111), no angular variation was observed in the ARUPS spectra of the monolayer. This indicates that there is no preferred orientation on the In_2O_3 surface. As was the case on Al(111), there are no differential shifts apparent in orbital binding energies between the monolayer and the multilayer. This implies that no particular orbital is involved in the bonding, and again only an electrostatic bond is indicated. The rigid shift of orbital emission energies between multilayer and monolayer, although nearly, are

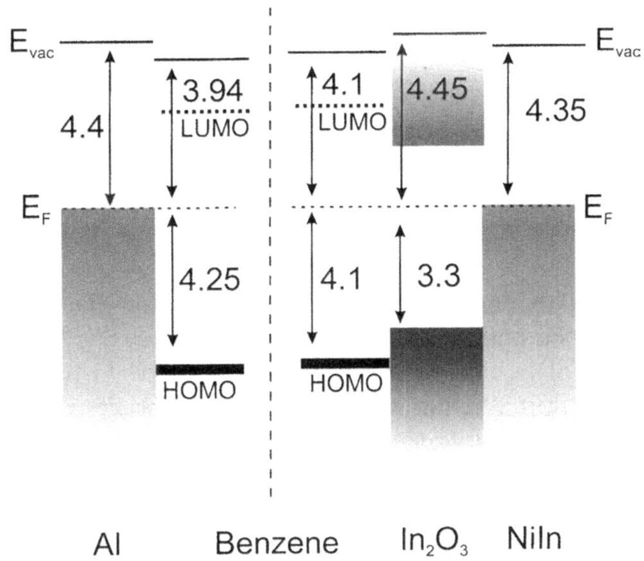

Fig. 5. Schematic diagram of the band alignment in an ideal Al/benzene/In_2O_3 device (all energies in eV)

not precisely equal to the changes in work function. This is presumably a final state effect in the photoemission due to poorer intermolecular screening of the photoholes in the multilayer films of these small oligomers.

Taking the multilayer HOMO binding energies from Figs. 3 and 4, together with that of the In_2O_3 valence band edge, and the work functions measured of the various surfaces, it is now possible to construct the band alignment diagram of an ideal Al/benzene/In_2O_3 device. This is shown in Fig. 5 where the *Fermi* levels of the two metals have been aligned, simulating a device under zero bias conditions. This device is clearly not optimal for efficient performance since the band alignment at both contacts is essentially the same. For an efficient device the LUMO should be close to the *Fermi* level at the electron injecting interface, whereas the HOMO should be close to the *Fermi* level at the hole injecting interface. By analogy, it would thus appear that an ideal, *i.e.* clean Al/sexiphenyl/*ITO* device should show poor performance. Indeed, investigations of prototype devices have shown that the cleaner Al interfaces result in poorer device performance [2]. It is known that oxygen induces a surface dipole on Al(111) which significantly reduces the work function. This results in a greater HOMO binding energy, thus bringing the LUMO closer to the *Fermi* level. In the Al/O/benzene case, the band alignment shifts linearly with the change in work function induced by oxygen at the Al interface [12]. It is important to note, however, that this is not a general case and only occurs here because the nature of the benzene bond (weak electrostatic) is not significantly altered by the oxygen at the interface. In general, the precise nature of the interfacial interaction will determine the band alignment. For example, for the Cu-thiophene oligomer interface the presence of oxygen has the opposite effect and leads to a much poorer band alignment because oxygen changes the interaction from a strong bond with significant charge transfer to a weaker electrostatic bond [13]. It is clear from these results that device manufacturers will need to have much greater control

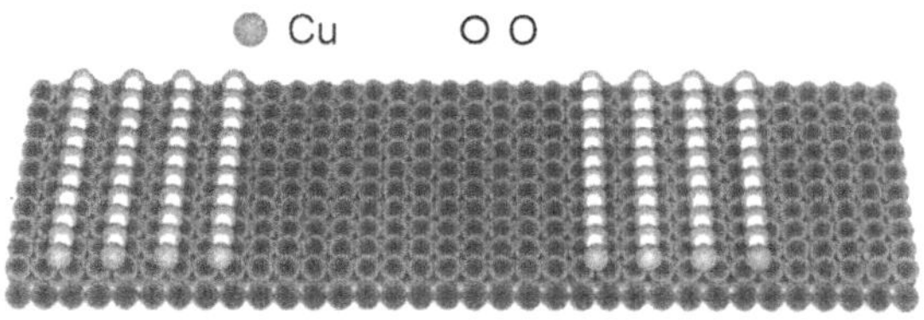

Fig. 6. Schematic drawing of the structure of the p(2×1)-O stripe phase on Cu(110)

of the oxygen partial pressures during electrode deposition than is presently the case if performance is to be optimized.

From the above it is clear that oxygen can be used to modify the electronic structure of metallic surfaces, but it can also serve to modify the surface geometry. Here we present preliminary results for the adsorption of bithiophene on the Cu(110)p(2×1)-O stripe phase, the structure of which is shown in Fig. 6. It consists of elongated Cu(110)p(2×1)-O islands, *i.e.* stripes separated by clean Cu(110) areas. By varying the oxygen coverage the width of the Cu-O stripes can be tuned between 20 and 100 Å [14, 15], whereas the periodicity is maintained. In this case we used approximately equal widths of clean Cu and Cu-O (periodicity 70 Å), on which bithiophene was adsorbed. Thermal desorption spectra of bithiophene from this surface are shown in Fig. 7. The spectrum of the multilayer film (Fig. 7a) shows two sharp features: the multilayer desorption at 225 K and the desorption of

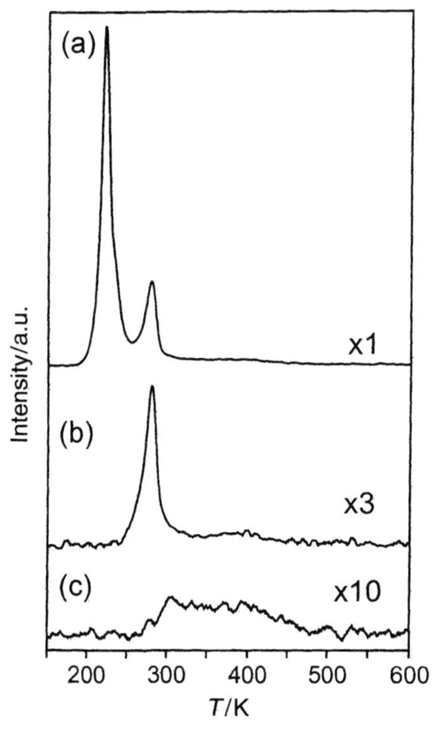

Fig. 7. Thermal desorption spectra of bithiophene on the Cu(110)p(2×1)-O stripe phase; (a) multilayer, (b) after heating to 240 K and cooling to 90 K, (c) after heating to 280 K and cooling to 90 K

bithiophene bonded to the CuO stripes at 280 K. A monolayer of bithiophene can be produced by heating a multilayer to 240 K and then allowing the sample to cool. A TDS spectrum from this surface is shown in Fig. 7b. Bithiophene desorption from the Cu-O stripes is still seen at 280 K, but desorption from the Cu stripes is also visible (note the multiplication factor) as a low, broad feature extending to 450 K. It is interesting to note that the presence of the CuO stripes actually stabilizes the bithiophene on the clean Cu stripes: on clean Cu(110) bithiophene desorption is complete at 350 K. Upon heating a bithiophene monolayer to 280 K and then cooling the bithiophene is desorbed from the Cu-O stripes, leaving the Cu stripes populated. This is evidenced by the spectrum in Fig. 7c which shows the broad feature characteristic of the Cu stripes with no desorption from Cu-O. The results of the adsorption of this model molecule show that the reconstruction does indeed provide the necessary differences in interface chemistry to allow for selective molecular adsorption and therefore a replication of the mesoscopic template pattern.

Although controlled surface science studies of active organic materials are still in their infancy, it is clear that they will play an important role in the development of organic device technology.

Experimental

The photoemission spectra were recorded using a Vacuum Generators ADES 400 spectrometer which was also equipped with a quadropole mass spectrometer for TDS measurements. The LED depth profile was measured with a Perkin-Elmer 610 *Auger* spectrometer. All measurements were performed in UHV. Benzene and bithiophene were purified by several freeze-pump-thaw cycles before dosing. Al(111) [12] and Cu(110)p(2×1)-O [14] surfaces were prepared according to the literature. The samples were maintained at 90 K during dosing.

Acknowledgements

We would like to thank the SFB *Elektroaktive Stoffe* of the Austrian Science Foundation for financial support and *Jörg Gomm* for invaluable technical work. The sexiphenyl LED was grown by the group of Prof. *Leising* at the Technische Universität Graz.

References

[1] Woodruff DP, Delchar TA (1994) Modern Techniques of Surface Science. Cambridge University Press
[2] Koch N, Pogantsch A, List EJW, Blyth RIR, Ramsey MG, Netzer FP, Leising G (1999) Appl Phys Lett **74**: 2909
[3] Blyth RIR, Ramsey MG, Resel R, Netzer FP (2000) Surf Sci **446**: 137
[4] Jaeger RM, Kuhlenbeck H, Freund HJ, Wuttig M, Hoffmann W, Franchy R, Ibach H (1991) Surf Sci **259**: 235
[5] Koller G, Netzer FP, Ramsey MG (1999) Surf Sci **421**: 353
[6] Netzer FP, Ramsey MG (1992) Crit Rev Sol State Mat Sci **17**: 397
[7] Ramsey MG, Steinmüller D, Netzer FP, Schedel T, Santaniello A, Lloyd DR (1991) Surf Sci **251**: 979
[8] Kimura K, Katsumumata S, Achiba Y, Yamazaki, Iwata S (1981) Handbook of He I Photoelectron Spectra of Fundamental Organic Molecules. Halsted, New York
[9] Netzer FP (1991) Langmuir **7**: 2544

[10] Duschek R, Mittendorfer F, Blyth RIR, Netzer FP, Hafner J, Ramsey MG (2000) Chem Phys Lett **318**: 43
[11] Barr TL, Liu YL (1989) J Phys Chem Solids **50**: 657
[12] Blyth RIR, Sardar SA, Netzer FP, Ramsey MG (2000) Appl Phys Lett, in press
[13] Koller G, Blyth RIR, Sardar A, Netzer FP, Ramsey MG (2000) Appl Phys Lett **76**: 927
[14] Zeppenfeld P, Diercks V, Tölkes C, David R, Krzyzowski MA (1998) Appl Surf Sci **130–132**: 484
[15] Kern K, Niehus H, Schatz A, Zeppenfeld P, George J, Comsa G (1991) Phys Rev Lett **67**: 855

Received May 30, 2000. Accepted September 21, 2000

Preparation and Radical Oligomerization of an Fe(II) Complex without Loss of Spin-Crossover Properties[a]

Gerald Schwarzenbacher[1], **Marion S. Gangl**[1], **Marian Goriup**[1], **Martin Winter**[2], **Matthias Grunert**[3], **Franz Renz**[3], **Wolfgang Linert**[3], and **Robert Saf**[1,*]

[1] Institute for Chemical Technology of Organic Materials, Technical University Graz, A-8010 Graz, Austria
[2] Institute for Chemical Technology of Inorganic Materials, Technical University Graz, A-8010 Graz, Austria
[3] Institute for Inorganic Chemistry, Vienna University of Technology, A-1060 Vienna, Austria

Summary. 11-(4*H*-1,2,4-Triazol-4-yl)-undecylmethacrylate (**1**), a new ligand for Fe(II) spin-crossover (SCO) complexes containing a polymerizable group, was synthesized and characterized. The complex $[Fe \cdot \mathbf{1}_3](BF_4)_2$ (**2**) was obtained by reaction of **1** with $Fe(BF_4)_2 \cdot 6H_2O$ (molar ratio **1**/Fe(II) = 3/1) in *THF*. Complex **2** showed a gradual spin-crossover between 80 and 230 K. The methacrylate units in the ligands of complex **2** could be oligomerized radically in solution (initiator: azoisobutyronitrile) without loss of the spin-crossover behaviour.

Keywords. Spin-crossover; High-spin; Low-spin; Triazole; Methacrylate; Oligomer; Polymer.

Introduction

Molecular magnetism – an area of research that includes experiments towards purely organic magnetic materials (*e.g.* low molecular weight organic radicals or polyradicals), metal-organic radical compounds, inorganic complexes, high-spin clusters, *etc.* – has attracted increasing attention in recent years [1, 2]. An important part of the corresponding activities is focused towards new materials exhibiting spin-crossover (SCO) behaviour [2–4]. These materials usually contain first row transition metal ions with an electronic configuration of d^4 to d^7 in octahedral surroundings, the ions existing either in a low-spin (LS) or in a high-spin (HS) ground state depending on the coordinated ligands. Under certain conditions, HS-LS transitions are possible. The actual configuration at ambient conditions depends, in a first approximation, on the energy gap Δ between the e_g and the t_{2g} orbitals of the metal ion and on the mean spin pairing energy P. For $P \gg \Delta$, as is the case in a

[a] Dedicated to Prof. Dr. *K. Hummel* on the occasion of his 70th birthday
* Corresponding author

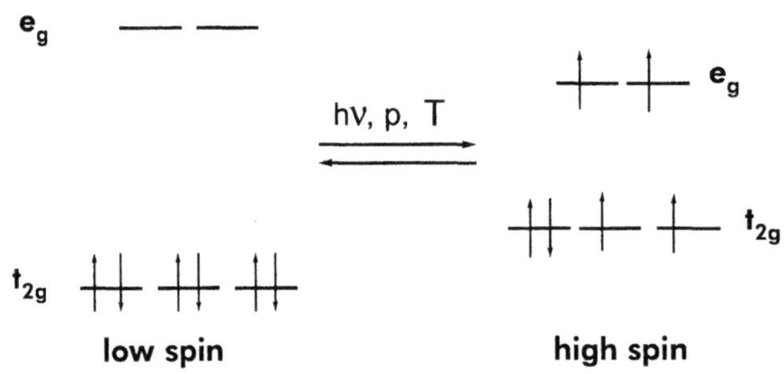

Fig. 1. Spin-crossover process in d^6-ions as *e.g.* iron(II)

rather weak ligand field, the complex exists in the HS state. In a rather strong ligand field ($P \ll \Delta$), the ground state is LS. If the conditions $P \gg \Delta$ or $P \ll \Delta$ are not fulfilled, LS-HS transitions are possible (Fig. 1). These processes can be induced by variation of temperature or pressure as well as by irradiation with light (LIESST or reverse LIESST; LIESST = Light Induced Excited Spin State Trapping) [5].

SCO induced by thermal energy or by irradiation with light are the most extensively investigated phenomena. Thermally induced spin crossing can occur very abruptly within a small thermal range (a few degrees) or gradual over a wide range, as one step transition or in two steps, with or without hysteresis [4]. Abrupt transitions in combination with hysteresis are typical for cooperative SCO. Independent of the individual types, the transitions are combined with drastic changes of the physical properties of the compounds, *e.g.* magnetic behaviour, bond lengths and, consequently, force constants, symmetry, or electronic structure. These changes offer many possibilities to characterize the spin-crossover phenomena. Important methods are for example magnetic measurements, electronic and *Mössbauer* spectroscopy, thermal analysis, IR and NMR spectroscopy [3].

Most experiments focused towards SCO deal with complexes of iron(II) [4–15]. However, spin-crossover complexes containing other metal ions, *e.g.* Mn^{3+}, Fe^{3+}, Co^{2+}, Co^{3+}, or Cr^{2+} are also known (Ref. [2] and references therein). As mentioned before, all ions with an electronic configuration d^4–d^7 are possible candidates. Concerning the ligands, a wide spectrum of different compounds has been investigated, especially nitrogen containing aromatic heterocycles. Typical examples are derivatives of phenanthroline [7], bipyridine [8], *bis*(benzimidazol-2-yl)pyridine [9], or tetrazole [10]. Experiments with derivatives of 1,2,4-triazole will be discussed below. Some SCO complexes with ligands containing sulfur or phosphor, respectively, are also known (*e.g.* dithiocarbamate with iron(III) or phoshonic acid dialkylester anions with cobalt(III) [2]).

Very important with respect to various applications is that the spin-crossover behaviour can be tuned by several factors. Especially the counter ions (coordinating or non-coordinating) and solvent molecules incorporated in the crystal lattice can be used to in- or decrease the transition temperature, or to influence the slope of the thermal transition curve and/or hysteresis. Furthermore, metal dilution, isotopic effects, and the size of the crystals can also change the transition behaviour. Thus, the preparation method of the complex can have remarkable effects on its SCO properties.

Their typical multifunctionality makes these materials interesting for various applications. Bistability, combined with light induced switchability, offers the possibility to use them for magnetooptical information storage devices, or even rewritable compact discs [11]. The change of the optical properties can be used within display devices [12]. The major requirements for these applications are a broad hysteresis of ideally about 50 K and, of course, a working temperature near ambient temperature. Furthermore, applications as molecular conductors, temperature/pressure and other sensors, *etc.* are in discussion.

As already mentioned, the interest in complexes exhibiting SCO behaviour has remarkably increased. However, there are only few investigations concerning spin-crossover compounds incorporated into organic macromolecular systems (*e.g.* solutions of SCO complexes in poly(vinyl alcolhol) have been investigated, especially to increase the order of the active centers by stretching of the polymer matrix [16]). In contrast to these experiments, the central aspect of the present work are experiments towards the covalent incorporation of spin-crossover complexes into organic polymeric matrices/networks. Successful covalent incorporation is interesting from various points of view. Phase separation between SCO complex and polymer matrix will be suppressed, and the preparation of new materials with a combination of properties typical both for the SCO complex and the organic polymer becomes feasible. Additionally, the approach offers new possibilities for the processing of SCO active materials, for example for the preparation of thin opto- or electroactive films.

Within this work we describe first experiments towards the covalent incorporation of an Fe(II)/triazole SCO complex into a polymer matrix by radical polymerization of methacrylate units attached to triazole ligands.

Results and Discussion

This work concentrates on SCO complexes based on derivatives of 4*H*-1,2,4-triazole (in the following text abbreviated as triazole). Triazoles substituted in position 4 are well known as suitable ligands *L* for Fe(II) spin-crossover complexes [6, 11, 13–15]. Depending on the chemical structure of the substituent and the molar ratio *L*/Fe(II), mono-, di-, tri-, or polynuclear compounds can be obtained [15]. Using 4-alkyl-triazoles and a molar ratio of $L/\mathrm{Fe(II)} = 3/1$ often results in the formation of polynuclear compounds with the composition $\mathrm{Fe}L_3A_{1\ \mathrm{or}\ 2} \cdot x$ solvent (*A*: *e.g.* Hal^-, BF_4^-, ClO_4^-, organosulfonates), where iron atoms are linked together *via* N1/N2 of three triazole ligands *L*. Materials exhibiting cooperative spin crossing at rather high temperatures can be obtained, even when the length of the alkyl chain is increased.

The preparation of a corresponding triazole with a polymerizable group at the end of a long alkyl chain seemed promising for us, since the distance between the groups involved in the polymerization reaction and the active spin-crossover center is rather high in this case. This should reduce negative effects of the polymerization on the spin-crossover complex, especially loss of the SCO behaviour. On the other hand, the flexibility of a long alkyl chain should facilitate polymerization.

Synthesis and characterization of ligand **1**

We chose the methacrylate as polymerizable group to investigate if SCO complexes based on triazoles can be oligomerized or polymerized radically without loss of

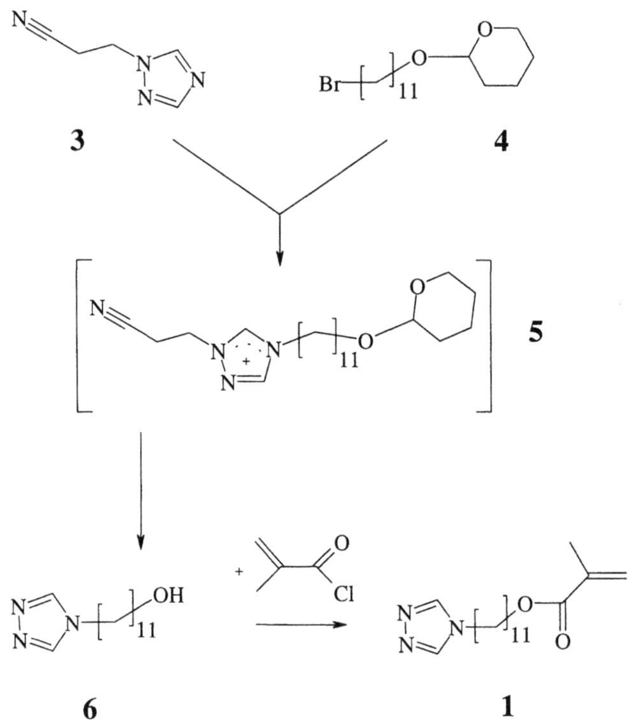

Scheme 1. Synthesis of 11-(4*H*-1,2,4-triazol-4-yl)-undecylmethacrylate (**1**)

their typical properties. The new ligand 11-(4*H*-1,2,4-triazol-4-yl)-undecylmethacrylate (**1**) was prepared according to Scheme 1.

1-Alkyl-1*H*-1,2,4-triazoles are the preferred product of the reaction of 1*H*-1,2,4-triazole with alkyl halides; 4-alkyl derivatives can be obtained when position 1 is protected. Thus, 1*H*-1,2,4-triazole-1-propanenitrile (**3**) was prepared as starting material by reaction of 1*H*-1,2,4-triazole with acrylonitrile [17]. The hydroxy group of 11-bromoundecanol was protected with dihydropyrane giving 2-(11-bromoundecoxy)-tetrahydropyrane (**4**) to avoid side reactions during the alkylation [18]. The reaction of triazole **3** with **4** was performed in nitromethane. The triazolium bromide **5** was not isolated. The protecting group at position 1 of triazole **5** was removed with sodium methylate. The alcohole function was recovered using hydrochloric acid to remove the *THP* protection group giving 4-(11-hydroxyundecyl)-4*H*-1,2,4-triazole (**6**). The new ligand **1** was obtained by esterification of alcohol **6** with methacroylchloride applying standard methods [19].

Additionally to standard characterization methods, electrospray ionization mass spectrometry (ESI-MS) was used to confirm the structure of compounds **6** and **1**. Figures 2a and 2b show the corresponding ESI mass spectra obtained in the high concentration regime (appr. 100 mg/dm^3 **6** or **1**, respectively). Both spectra exhibit the expected *pseudo*-molecular and cluster ions $[M \cdot Cs]^+$, $[2M \cdot Cs]^+$, and $[3M \cdot Cs]^+$, respectively, and incomplete desolvated ions (*e.g.* $[M \cdot ac \cdot Cs]^+$, $ac =$ acetone). The ion $[\mathbf{6} \cdot Cs]^+$ was found at $m/z_{\text{observed}} = 372.12$ Da (theoretical value: $m/z_{\text{theor}} = 372.10$ Da), $[\mathbf{1} \cdot Cs]^+$ at $m/z_{\text{observed}} = 440.15$ Da ($m/z_{\text{theor}} = 440.13$ Da). Only the *pseudo*-molecular ions $[M \cdot Cs]^+$ were found when the samples were

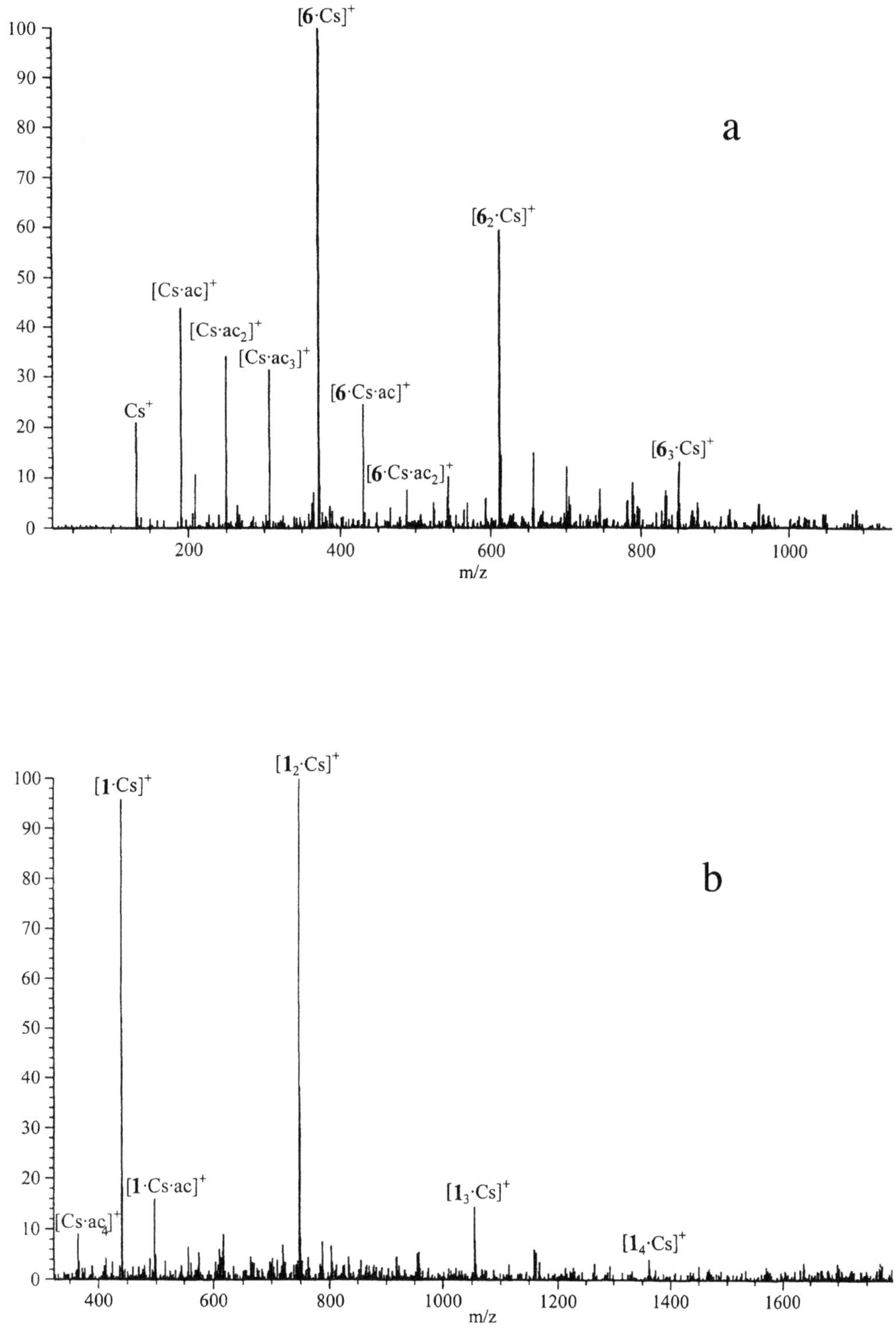

Fig. 2. ESI mass spectra of alcohol **6** (a) and methacrylate **1** (b) (*ac* = acetone)

diluted to a concentration of 1 mg/dm^3. Together with the NMR data (see Experimental) this ensures that no partial di- or trimerization of methacrylate **1** occured during preparation or workup.

Synthesis and characterization of complexes **2** *and P*-**2**

In principal, radical polymerizations can be carried out in many different ways, *e.g.* in solution or in bulk, with various initiators. Within this work, polymerization in solution was investigated which requires a soluble SCO complex. By screening experiments it was found that the complex prepared from ligand **1** and $Fe(BF_4)_2 \cdot 6H_2O$ (molar ratio **1**/Fe(II) = 3/1) is soluble in *THF*, a solvent suitable for radical polymerization reactions. In other solvents, *e.g.* methanol, rapid precipitation of a white powder occurred. Being beyond the scope of this work, these materials were not further investigated.

Azoisobutyronitrile (*AIBN*) was used as an initiator to minimize interactions with the Fe(II) complex (*e.g.* peroxides as benzoyl peroxide might oxidize Fe(II); peroxodisulfates might additionally promote the exchange of anions).

It is well known that the magnetic properties of SCO complexes can be very sensitive to minor differences during the preparation, especially when triazoles are used as ligands [15]. Thus, the following experimental sequence was applied:

(*i*) Preparation of the soluble SCO complex by reaction of ligand **1** with $Fe(BF_4)_2 \cdot 6H_2O$ (molar ratio **1**/Fe(II) = 3/1) in *THF*, followed by division of the solution into two parts of equal volume (samples *A* and *B*, respectively). This ensured that the complexes with monomeric (**2**) and polymerized (*P*-**2**) ligands were prepared from solutions of exactly the same ratio **1**/Fe(II).
(*ii*) Addition of an appropriate amount of *AIBN* to sample *B* followed by heating of both samples to reflux for 10 h. Polymerization of the methacrylate units occurred in sample *B*. Sample *A* was refluxed to ensure that both samples experienced the same thermal history.
(*iii*) Cooling of both samples to room temperature followed by storage under argon without stirring for 24 h.
(*iv*) Precipitation of samples by pouring into a suitable poor solvent (*n*-pentane).
(*v*) Drying in vacuum to achieve complete removal of solvents.

Figure 3 shows the IR spectra of complexes **2** and *P*-**2**. The IR spectrum of **2** proved that the methacrylate units of the ligands were not affected during the formation of the complex, *e.g.* by partial redox polymerization. The intensity of the C=O absorption (1717 cm^{-1}) in relation to the intensity of the C=C absorption (1637 cm^{-1}) was the same for **1** and **2**. A distinct difference was observed for the typical triazole N–C=N band. For ligand **1** this band was observed at 1530 cm^{-1}. The IR spectrum of complex **2** showed the corresponding absorption at 1550 cm^{-1}. The quantitative shift of the sharp band – together with the absence of other new bands in this region – implies that only one type of coordination of triazol to Fe(II) exists in **2**. In analogy to Fe(II)/4-alkyl-triazole complexes described in the literature, a polynuclear structure where iron atoms are linked together via N1/N2 of three triazole groups is expected [15]. The absence of solvent molecules (thermal analyses up to 250°C) implies the composition $[Fe \cdot \mathbf{1}_3](BF_4)_2$ for complex **2**. This

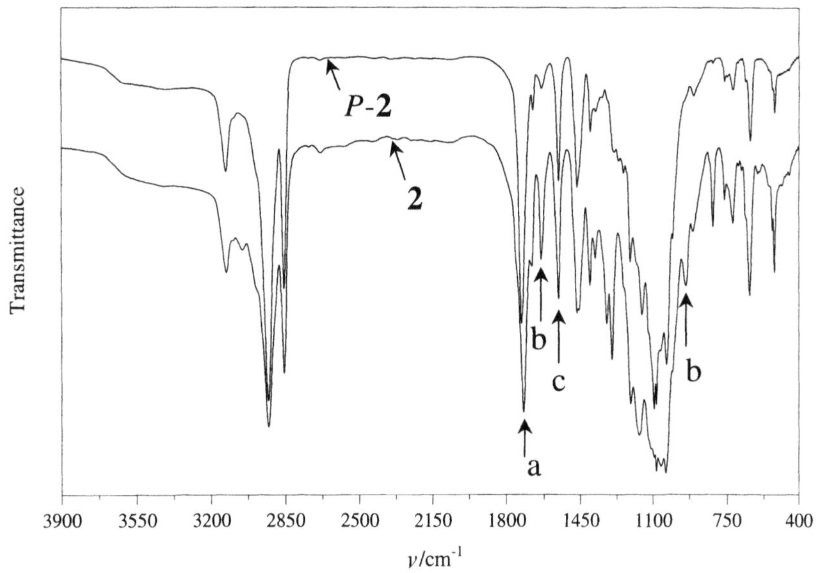

Fig. 3. IR spectra of **2** and *P*-**2**; characteristic absorptions of carbonyl group (a), methacrylic double bond (b), and triazole (c)

structure is in accordance with results obtained from magnetic measurements (see below).

The IR spectrum of *P*-**2** showed significant differences. First of all, compared to the IR spectrum of **2**, the intensity of absorptions typical for the methacrylic double bond (1637 cm^{-1} and 944 cm^{-1}) was reduced. The amount of methacrylate units remaining unreacted is estimated to less than 10%. Additionally, the maximum of the C=O absorption shifted from 1717 cm^{-1} to 1729 cm^{-1}, as it has to be expected due to the conversion of the majority of double bonds conjugated to C=O. These changes prove successful polymerization. On the other hand, no significant changes were observed near the typical triazole band at 1550 cm^{-1} (see discussion above). This implies that the center of the complex was not influenced by the polymerization. Consequently, the materials **2** and *P*-**2** should have comparable magnetic properties. Figure 4 shows the product $\chi_M \cdot T$ *vs.* T for both complexes. The susceptibility was measured in steps of 2 K. In order to reach thermal equilibrium, the temperature oscillated around the fixed value for a few minutes. Complex **2** showed a rather gradual transition from LS to HS in the thermal range of 80–250 K, which was accompanied by a change of colour from violet to slightly yellow. It has to be mentioned that this gradual transition differs significantly from what has been observed for various other Fe(II)/triazole polynuclear complexes. For most of the corresponding materials, the spin crossing is complete within a much smaller thermal range. The difference can be explained by the method how complex **2** was prepared. The precipitation was achieved by pouring a solution of the complex in *THF* into *n*-pentane; thus, a product with a reduced degree of order should be obtained, especially compared to materials prepared by slow crystallization. Nevertheless, the results of the magnetic measurements show that spin-crossover occurs for a rather high percentage of Fe(II), $\chi_M \cdot T$ increasing from 1 $cm^3 \cdot K/mol$

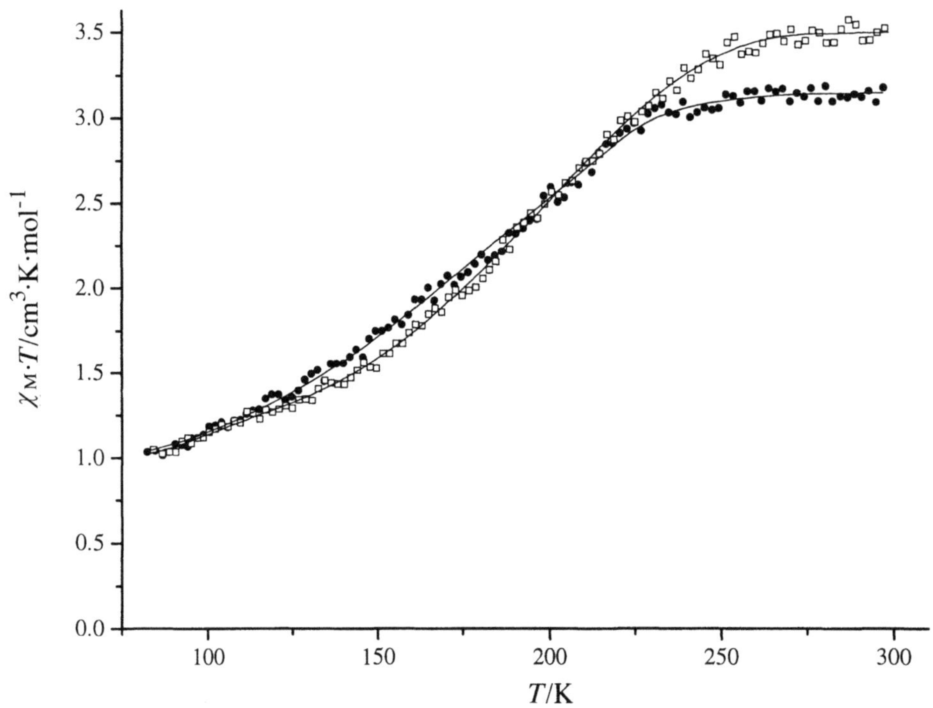

Fig. 4. Magnetic properties of **2** (•) and *P*-**2** (□)

at 80 K to 3.15 cm^3 · K/mol at room temperature. Complex *P*-**2** showed comparable but not absolutely identical magnetic properties. The value of $\chi_M \cdot T$ at room temperature increased to 3.5 cm^3 · K/mol, a result that corresponds well to what is expected for the high-spin state. When the temperature was reduced to 80 K, $\chi_M \cdot T$ again decreased to approximately 1 cm^3 · K/mol. Compared to product **2**, the spin-crossover seems to occur a little bit sharper. As described in the experimental section, the precipitation of *P*-**2** started 1 h after *AIBN* was added to the solution of complex **2**. Heating to reflux, however, was continued for additional 9 h to complete the polymerization of methacrylate units. During this time already precipitated material remained in a swollen state. Thus, partial ordering processes should still be possible, and disordering processes should be reduced in comparison to the completely dissolved state. Thus, a slightly increased degree of order in *P*-**2** is expected and serves as a rationale for the deviation of the magnetic properties. As already mentioned, at a temperature of 80 K $\chi_M \cdot T$ was still approx. 1 cm^3 · K/mol for both **2** and *P*-**2**, respectively. This can be explained by incompleteness of the gradual transition (see slope of both curves shown in Fig. 4 at low temperature) and, analogous to other Fe(II)/triazole polynuclear complexes described in the literature [15], by assuming permanently high-spin Fe(II) atoms at the ends of the polynuclear chains. However, the important aspect of these results is that complex **2** could be polymerized successfully without loss of the SCO behaviour.

The molecular weight distribution of the polymerized ligands of *P*-**2** was determined by means of size exclusion chromatography (SEC). Complex *P*-**2** was destroyed with HCl (2 mol/dm^3); subsequently, NaOH was added to precipitate iron as its hydroxide. The dispersion was extracted several times with $CHCl_3$. The organic layers were evaporated, and the residue was investigated by SEC. The maximum of the molecular weight distribution was found at 2070 g/mol (*vs.* polystyrene). This corresponds to 6–7 monomeric units. The low degree of polymerization is due to the relative high molar ratio of *AIBN*/methacrylate units (0.09).

The result clearly demonstrate that Fe(II)/triazole spin-crossover complexes containing methacrylic units in the ligands can be polymerized radically without destroying the spin-crossover properties. It is important to note that even at the relatively high molar ratio of *AIBN*/methacrylate units used within this work no negative influence of the radical polymerization on the SCO centers was observed. The strategy might generally be used for the covalent and homogeneous incorporation of SCO active complexes into polymer matrices, *e.g.* for the preparation of multifunctional transparent films.

Experimental

General

All commercial substances (Fluka, Aldrich) were used as purchased. The starting materials 1*H*-1,2,4-triazole-1-propanenitrile (**3**) and 2-(11-bromoundecoxy)-tetrahydropyrane (**4**) were synthesized as described in the literature [17, 18]. Solvents were dried according to standard procedures [20]. ^1H and ^{13}C NMR spectra were recorded on a Bruker MSL 300 NMR spectrometer. IR measurements were performed from KBr pellets on a Perkin Elmer Spectrum One FTIR spectrometer. Thermal analyses were carried out on a STA 625 from Rheometrics, elemental analyses on a Carlo Erba 1106 instrument. The magnetic properties were measured on a Lake Shore 7221 AC-susceptometer/DC-magnetometer. The results were corrected for the magnetization of the sample holder and for diamagnetic contributions, which were estimated from *Pascal* constants. ESI mass spectra were recorded on a Kratos Profile double-focusing mass spectrometer using a Kratos standard electrospray ion source. A mixture of acetone/H_2O (99/1) containing CsI (50 mg/dm^3) was used as solvent; details of the method will be described elsewhere [21].

Synthesis of ligand **1**

4-(11-Hydroxyundecyl)-4H-1,2,4-triazole (**6**; $C_{13}H_{25}N_3O$)

1.6 g (13 mmol) of triazole **3** and 14 g (12 mmol) of bromide **4** were dissolved in 8 cm^3 of nitromethane. 100 mg of KI were added, and the solution was refluxed for 4 days. After cooling to room temperature, 11 cm^3 of sodium methylate (1.5 mol/dm^3 in MeOH) were added. The reaction mixture was stirred for 20 min; then, 22 cm^3 of NaOH solution (10% in H_2O) were added. The solution was extracted three times with 20 cm^3 of $CHCl_3$. The organic fractions were combined, and the solvent was removed by evaporation under reduced pressure. The residue was dissolved in 30 cm^3 of MeOH. After addition of 20 cm^3 of H_2O and 24 cm^3 of conc. HCl the solution was refluxed for 2 h and subsequently cooled to room temperature. After addition of 11 g of NaOH dissolved in 60 cm^3 of H_2O the mixture was extracted three times with 80 cm^3 of $CHCl_3$. The organic fractions were combined, dried over Na_2SO_4, and the solvent was removed by evaporation under reduced pressure. The crude product was purified by column chromatography (eluent: MeOH/$CHCl_3$ = 1/9) to yield 2.4 g (10 mmol, 89%) of **6**.

^1H NMR ($CDCl_3$, 300 MHz, δ) 1.3–1.8 (alkyl chain, m, 18H), 3.0 (OH, s, 1H), 3.6 (CH_2–OH, t, 2H), 4.0 (CH_2-triazole, t, 2H), 8.15 (triazole-H, s, 2H) ppm; ^{13}C NMR ($CDCl_3$, 75 MHz, δ): 25.8–32.8

(alkyl chain), 45.4 (CH_2-triazole), 62.7 (CH_2–OH), 142.8 (triazole-C) ppm; IR: $\nu = 3325$ (OH), 3115 (triazole), 2925 and 2845 (C–H), 1680, 1532 (triazole), 1457 (CH_2), 1175 (triazole), 1071 (C–O), 720 (CH), 630 cm^{-1}.

11-(4H-1,2,4-Triazol-4-yl)-undecylmethacrylate (**1**; $C_{17}H_{29}N_3O_2$)

3.0 g (12.5 mmol) of alcohol **6** were dissolved in 20 cm^3 of dry pyridine under argon, and the solution was cooled to 0°C. 1.84 cm^3 (18.8 mmol) of methacroylchloride were added dropwise keeping the temperature below 5°C. After stirring of the reaction mixture at 0–5°C for 1 h the ice bath was removed; the mixture was allowed to warm to room temperature, and stirring was continued for further 1.5 h. The residue obtained by evaporation of the solvent under reduced pressure was dispersed in 150 cm^3 of $NaHCO_3$ (5% in H_2O) and stirred for 20 min. The mixture was extracted three times with 150 cm^3 of $CHCl_3$. The organic fractions were combined, dried over Na_2SO_4, and evaporated. The crude product was purified by column chromatography (eluent: $MeOH/CHCl_3 = 1/9$) to yield 2.35 g (7.6 mmol, 61%) **1**.

^{1}H NMR ($CDCl_3$, 300 MHz, δ): 1.1–1.7 (alkyl chain, m, 18H), 1.9 (CH_3, s, 3H), 4.0 (CH_2-triazole, t, 2H), 4.1 (CH_2-ester, t, 2H), 5.5 (C=C–H, s, 1H), 6.15 (C=C–H, s, 1H), 8.15 (triazole-H, s, 2H) ppm; ^{13}C NMR ($CDCl_3$, 75 MHz, δ): 24.9–30.8 (alkyl chain, CH_3), 45.4 (CH_2-triazole), 64.8 (CH_2-ester), 125.2 and 136.53 (C=C), 142.8 (triazole-C), 167.57 (C=O) ppm; IR: $\nu = 3112$ (triazole), 2928 and 2855 (C–H), 1717 (C=O), 1677, 1637 (C=C), 1530 (triazole), 1455 (CH_2), 1296, 1180 (CO), 1037, 944 (C=C), 638 cm^{-1}.

Synthesis of complex **2** *and polymerization to P*-**2**

1.00 g (3.26 mmol) of ligand **1** were dissolved in 20 cm^3 of *THF* under argon. After addition of 367 mg (1.09 mmol) of $Fe(BF_4)_2 \cdot 6H_2O$ dissolved in 10 cm^3 of *THF* the solution was refluxed for 15 min. The yellow solution was concentrated to a total volume of approx. 10 cm^3, cooled to room temperature, and stored under argon without stirring for 24 h. No precipitation or change in colour occurred during this time. Subsequently the solution was divided into two parts of equal volume (samples *A* and *B*).

25 mg of azoisobutyronitrile (*AIBN*) were added to sample *B* and both samples were refluxed for 10 h. Sample *A* remained a yellow solution, whereas precipitation started in sample *B* after *ca.* 1 h. At the end of the reaction time, the colour of sample *B* had changed to orange. Both samples were allowed to cool to room temperature. After 24 h without stirring the samples were poured into 20 cm^3 of *n*-pentane. The precipitates were filtered off, washed twice with 10 cm^3 of *n*-pentane, and dried *in vacuo*.

83.8 mg of white **2** were obtained from sample *A*, 89.9 mg of slightly yellow *P*-**2** from sample *B*.

2: IR: $\nu = 3125$ (triazole), 2928 and 2854 (C–H), 1717 (C=O), 1679, 1637 (C=C), 1550 (triazole), 1455 (CH_2), 1296, 1181 (CO), 1035, 944 (C=C), 640 cm^{-1}; $C_{51}H_{87}N_9O_6Fe(BF_4)_2$ (1151.78); calcd.: C 53.2, H 7.6, N 11.0; found: C 52.9, H 7.4, N 11.2.

P-**2**: IR: $\nu = 3127$ (triazole), 2928 and 2855 (C–H), 1729 (C=O), 1677, 1551 (triazole), 1458 (CH_2), 1175 (CO), 1084, 1033, 638 cm^{-1}; $C_{51}H_{87}N_9O_6Fe(BF_4)_2$ (1151.78); calcd.: C 53.2, H 7.6, N 11.0; found: C 53.4, H 7.5, N 10.8.

Acknowledgements

The authors wish to thank *Renate Landgraf* and *Robert Rüf* for participation in the experimental work. Financial support by the Austrian Science Foundation (FWF) (SFB Electroactive Materials), the European Community (Training and Mobility of Researchers Network (TMR), project "Thermal and Optical Spin State Switching" (TOSS), contract number ERB-FMRX-CT98-0199), and the European Science Foundation (project "Molecular Magnets") is gratefully acknowledged.

References

[1] Lahti PM (1999) Magnetic Properties of Organic Materials. Dekker, New York
[2] Kahn O (1993) Molecular Magnetism. VCH, Weinheim, New York
[3] Gütlich P (1981) Struct Bonding (Berlin) **44**: 83
[4] Gütlich P, Jung J (1995) J Mol Struct **347**: 21
[5] Gütlich P (1997) Mol Cryst Liq Cryst **305**: 17
[6] Kahn O, Codjovi E (1996) Phil Trans R Soc Lond A **354**: 359
[7] Sugiyarto KH, Craig DC, Goodwin HA (1996) Aust J Chem **49**: 497
[8] Soyer H, Mingotaud C, Boillot ML, Delhaes P (1998) Langmuir **14(20)**: 5890
[9] Boca R, Baran P, Dlhan L, Fuess H, Haase W, Renz F, Linert W, Svoboda I, Werner R (1997) Inorg Chim Acta **260**: 129
[10] Poganiuch P, Decurtins S, Gütlich P (1990) J Am Chem Soc **112(9)**: 3270
[11] Kahn O, Martinez CJ (1998) Science **279**: 44
[12] Kahn O, Codjovi E, Garcia Y, van Koningsbruggen PJ, Lapouyade R, Sommier L (1996) Spin-Transition Molecular Materials for Display and Data Processing. In: Molecule Based Magnetic Materials Symp Ser No 644, Turnbull, p 298
[13] Garcia Y, van Koningsbruggen PJ, Lapouyade R, Fournès L, Rabardel L, Kahn O, Ksenofontov V, Levchenko G, Gütlich P (1998) Chem Mater **10**: 2426
[14] Verelst M, Sommier L, Lecante P, Mosset A, Kahn O (1998) Chem Mater **10**: 980
[15] Haasnoot JG (1996) 1,2,4-Triazoles as ligands for iron(II) high spin–low spin crossovers. In: Kahn O (ed) Magnetism: A Supramolecular Function, Kluwer, p 299
[16] Renz F, Hasegawa M, Hoshi T, El-Ayaan U, Linert W, Fukuda Y (1999) Mol Cryst Liq Cryst **335**: 531
[17] Horváth A (1995) Synthesis **9**: 1183
[18] Baldo MA, Chessa G, Marangoni G, Pitteri B (1987) Synthesis **8**: 720
[19] Becker HGO et al (1993) Organikum. 19. Aufl, Barth, Leipzig Berlin Heidelberg
[20] Perrin DD, Armarego WLF, Perrin DR (1980) Purification of Laboratory Chemicals. Pergamon Press, Oxford Frankfurt
[21] Schwarzenbacher G, Mirtl C, Saf R, in preparation

Received May 30, 2000. Accepted December 10, 2000

Phase Gratings in Photoreactive Polymers: A Way to Optically Pumped Organic Lasers[a]

Thomas Kavc[1], **Wolfgang Kern**[1,*], **Christian Zenz**[2], **Günther Leising**[2], **Gerald Kranzelbinder**[3], and **Eric Toussaere**[3]

[1] Institut für Chemische Technologie Organischer Stoffe, Technical University of Graz, A-8010 Graz, Austria
[2] Institut für Festkörperphysik, Technical University of Graz, A-8010 Graz, Austria
[3] Ecole Normale Superieure de Cachan, LPQM, F-94230 Cachan, France

Summary. Poly-(4-vinylbenzyl-thiocyanate) (*PVBT*) was investigated as a photoreactive polymer which changes its refractive index upon UV irradiation as is required for the production of phase gratings. After UV irradiation, the refractive index of *PVBT* films increased from $n_D^{20} = 1.629$ to $n_D^{20} = 1.660$ ($\Delta n = +0.031$). The change in the refractive index resulted mainly from the photoinduced isomerization of thiocyanate (SCN) to isothiocyanate (NCS) groups. It was possible to modify irradiated *PVBT* with 1-naphthylmethyl amine which lead to a further increase of the refractive index. Both the isomerization and the modification were followed with FTIR spectroscopy. This modification of the irradiated zones also yields an advantageous stabilization of phase gratings enscribed in *PVBT*. The investigations were carried out in the realm of novel organic distributed feedback (DFB) lasers.

Keywords. Refractive index; Phase grating; Photoisomerization; Thiocyanate; Distributed feedback; Organic laser.

Introduction

Organic conjugated materials are presently attracting a lot of interest due to applications in light emitting diodes (LEDs) [1], light emitting electrochemical cells [2], photodetectors [3], field effect transistors [4], and solar cells [5]. Moreover, the progress in the chemical synthesis of highly luminescent conjugated polymers resulted in a novel class of solid state laser materials [6]. The excellent intrinsic optical properties make organic materials also potential candidates for laser applications. Optically pumped organic solid state lasers based on conjugated polymers, oligomers, and charge transfer systems [7–10] have already been demonstrated (for an overview, see *e.g.* Ref. [11]).

Conjugated polymers in solution exhibit high intrinsic emissive properties which are preserved also in the solid state. Thus, high photoluminescence quantum

---

[a] Dedicated to Prof. *K. Hummel* (Graz) on the occasion of his 70th birthday
* Corresponding author

yields (Φ_{lum}) up to 62% [12] and high cross sections for stimulated emission (up to 10^{-15} cm^{-2} [13–14]) have been observed in neat films of conjugated polymers. Further improvements in materials and designs are required for the development of electrically pumped organic laser diodes. Therefore, the development of solid-state lasers as well as other lasing applications employing conjugated polymers as the active material are of immediate research interest.

Organic materials are usually prepared as thin films by evaporation or spin casting procedures to give thicknesses in the micrometer range. Simple dielectric slab waveguide geometries can be obtained when the refractive index of the organic material is higher than that of the surrounding media. To realize laser structures, the introduction of optical feedback employing microcavities [7] or distributed feedback [15] are most promising. As has been shown by *Kogelnik* and *Shank* [16], optical feedback in a waveguide can be introduced by a periodical spatial variation of the refractive index (or the optical gain) of the active medium. In this case, optical feedback is provided by backward scattering of the guided wave along its optical path. In both cases, the laser emission wavelength λ is determined by the grating period Λ_p of the resonator and the effective refractive index n_{eff} of the structure according to the *Bragg* condition $\lambda = 2 \cdot n_{eff} \cdot \Lambda_p \cdot m^{-1}$, where m is the order of feedback. This concept of distributed feedback (DFB) has been successfully applied to films of conducting homopolymers such as *PPV* derivatives cast onto SiO_2 [8] or ladder-type poly-(paraphenylene) (*m-LPPP*) cast onto flexible plastic substrates [17]. In these cases, relief gratings were produced on the surface of the substrate to give grating periods of $\Lambda_p = 170$ and 300 nm, respectively. In another approach, in-plane distributed *Bragg* reflectors (DBR) can be employed which use the modulation of the refractive index to form an outer zone laser cavity. Figure 1 shows the schematical setup of DFB and DBR structures [15].

The present paper aims at the construction of a new type of grating. Instead of patterning the interface between the substrate and the active layer, the active layer itself is patterned. This can be achieved by a periodical change of the refractive index of the active material along the laser cavity. A periodical variation of the refractive index can be achieved by lithographic techniques which are capable of series production.

Currently, the modulation of the refractive index of polymers is utilized for holographic recordings [18]. As an example, poly-(vinylcinnamate) undergoes a

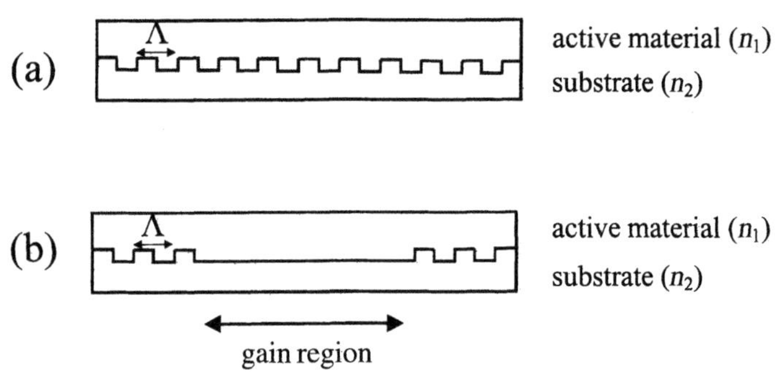

Fig. 1. Schematic setup of (a) distributed feedback lasers (DFB) and (b) in-plane distributed *Bragg* reflectors (DBR); $n_2 < n_1$

[2 + 2] cycloaddition under UV irradiation (365 nm). The photoreaction causes a reduction of the refractive index by *ca.* 0.02 [19]. As a further example, silicone polymers with dithienylethylene moieties along the main chain were reported to give large refractive index changes (Δn *ca.* 0.04) upon UV irradiation [20]. Alternatively, gratings in polymer films are obtained by UV-induced polymerization (*e.g.* acrylate resins [18]) and by UV-induced crosslinking (*e.g.* gelatine/dichromate [18, 21]). In another approach, matrix polymers insensitive towards radiation have been doped with photosensitive low molecular weight compounds to obtain changes of the refractive index [22, 23]. Such modifications of the refractive index of organic materials may also be useful as a simple technique for lateral definition of waveguides in integrated optics [24].

Gratings may be enscribed into polymer films by lithographic techniques using standard photomasks (Cr on glass). This technique as well as holographic lithography give a resolution around the wavelength of irradiation. With projection lithography based on off-axis illumination, even smaller grating periods (near $\lambda/2$) can be resolved. Consequently, for high spatial resolution the use of short UV wavelengths is suggested. Although excimer laser lithography with $\lambda = 248$ nm (KrF) is well established, there is a lack of holographic recording materials which operate in this UV range and at VUV wavelengths ($\lambda = 193$ (ArF) and 157 (F_2) nm).

In our recent works we have described the photoreactions of polymers containing photolabile pseudohalide substituents (cyanate (OCN), thiocyanate (SCN), and selenocyanate (SeCN)) [25–28]. Upon irradiation, these substituents undergo a photoisomerization to give the corresponding isocompounds (NCO, NCS, and NCSe). The present work is concerned with the refractive index changes which occur in poly-(4-vinylbenzyl-thiocyanate) (*PVBT*) during UV irradiation. The photochemical enscribing of phase gratings into this polymer is regarded and a method to stabilize the refractive index pattern by subsequent derivatization is also described. A stabilization of the pattern is of general importance, since photochemically enscribed phase gratings may be wiped out during the operation of a laser device. The paper is a contribution towards further development of organic laser devices.

Results and Discussion

Photochemical conversion of thiocyanates to isothiocyanates

The photoisomerization of benzyl thiocyanate to benzyl isothiocyanate was first described by *Parks* and *Spurlock* [31]. The reaction is practically irreversible; the photoequilibrium for the benzyl thiocyanate – benzyl isothiocyanate system lies almost completely on the side of the isothiocyanate (99:1 in acetonitrile solution [31]). A similar reaction occurs in benzyl selenocyanate [26] and also in derivatives of phenyl cyanate [25]. The photoreactions proceed *via* radical intermediates as shown for benzyl thiocyanate in Fig. 2. Our present investigation is restricted to esters of thiocyanic acid, as these compounds are more stable than organic cyanates and selenocyanates.

Generally, the refractive index of transparent organic compounds may be computed with the well-known *Lorentz-Lorenz* equation where R is the molar

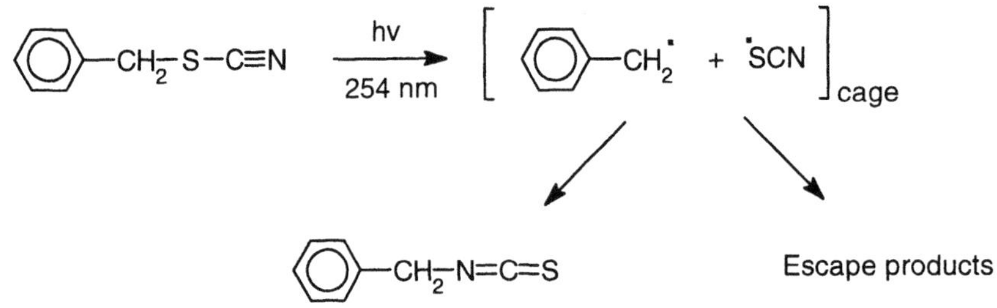

Fig. 2. Photoisomerization of benzyl thiocyanate

refraction (in $cm^3 \cdot mol^{-1}$), M is the molar mass (in $g \cdot mol^{-1}$), and ρ is the density (in $g \cdot cm^{-3}$).

$$R = \frac{n^2 - 1}{n^2 + 2} \cdot \frac{M}{\rho}$$

For most organic compounds, the molar refraction R, which is largely independent of temperature and physical state, can be calculated by summation of atomic and structural constants (see *e.g.* the tables published by *A. I. Vogel* [32]). The refractive index of a given compound then depends on the molar refraction and on the density. Considering the molar refractions of benzyl thiocyanate ($R_{calc} = 43.41\,cm^3 \cdot mol^{-1}$) and benzyl isothiocyanate ($R_{calc} = 45.63\,cm^3 \cdot mol^{-1}$), the iso-compound is expected to have a higher index of refraction. We measured the refractive index of benzyl thiocyanate and benzyl isocyanate at 50°C and obtained $n_D^{50} = 1.5620$ for benzyl thiocyanate and $n_D^{50} = 1.5920$ for benzyl isothiocyanate. Although data on the density of these compounds at 50°C are not available, it is reasonable to assume that ρ does not change significantly as the SCN groups are converted into NCS units. This assumption is justified by a comparison of the densities of aliphatic thiocyanates and the corresponding isothiocyanates (difference usually below 1%) [33]. From this it is apparent that the difference Δn between the refractive indices of thiocyanates and the corresponding isothiocyanates is mainly caused by their different molar refraction R.

Photoreactions in poly-(4-vinylbenzyl-thiocyanate)

From the data obtained so far, we expected polymers bearing pseudohalide substituents such as SCN to give pronounced changes of the refractive index upon irradiation. We investigated poly-(4-vinylbenzyl-thiocyanate) (*PVBT*), which was obtained from the monomer 4-vinylbenzyl-thiocyanate by free radical polymerization [26]. The film forming properties of this polymer are good, and films with thicknesses between 80 nm and several micrometers were easily obtained by spincasting techniques. *PVBT* shows an UV absorption peak around $\lambda = 233$ nm and is fully transparent at $\lambda > 290$ nm.

The UV irradiation of *PVBT* films with 254 nm light leads to a photoisomerization of the SCN units and to concomitant photo-crosslinking. The isomerization reaction was readily monitored by means of FTIR spectroscopy. The sharp SCN band at $2150\,cm^{-1}$ decreased, and at the same time broad signals typical of NCS units evolved around 2085 and $2170\,cm^{-1}$ (see insert in Fig. 3).

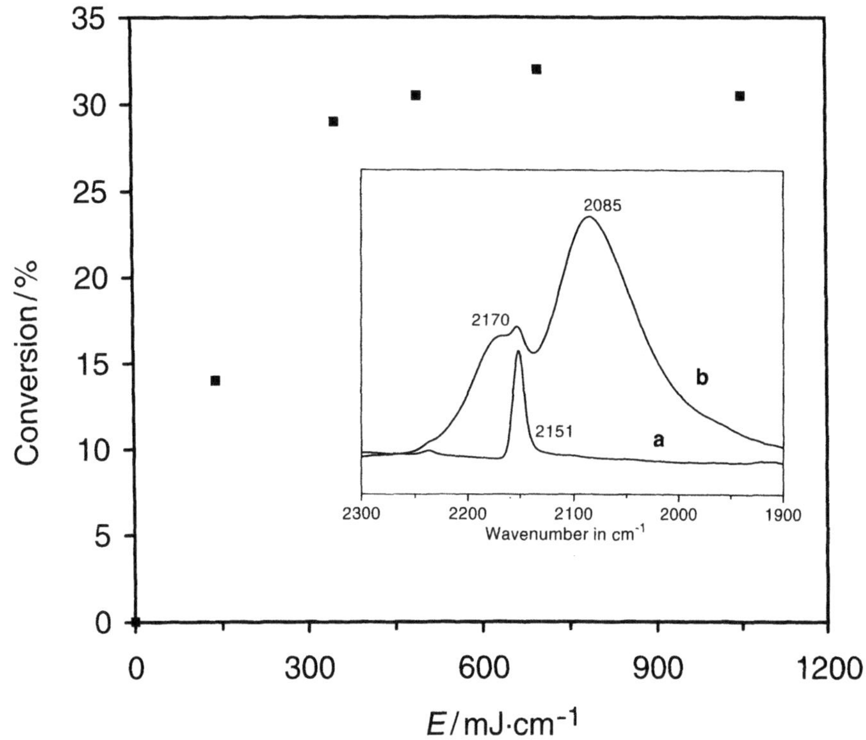

Fig. 3. *PVBT* film (thickness: 400 nm) under 254 nm UV irradiation; conversion of SCN units into NCS units as a function of the incident energy density *E*; insert: FTIR spectra of *PVBT* (a) before and (b) after irradiation with $E = 1000\,\mathrm{mJ \cdot cm^{-2}}$

For a more quantitative evaluation the infrared absorbance coefficients of the SCN and NCS groups in low molecular weight model compounds were determined. The SCN band in benzyl thiocyanate (in cyclohexane solution) gave $\varepsilon_{2158} = 75\,\mathrm{dm^3 \cdot mol^{-1} \cdot cm^{-1}}$, whereas for the NCS band in benzyl isothiocyanate (in cyclohexane solution) $\varepsilon_{2073} = 600\,\mathrm{dm^3 \cdot mol^{-1} \cdot cm^{-1}}$ was obtained. From absorbance FTIR spectra of irradiated *PVBT* films the photoisomerization was evaluated quantitatively. The progress of the photoisomerization is shown in Fig. 3, which displays the fraction of SCN units converted into NCS groups as a function of the incident energy density *E* at $\lambda = 254$ nm. All data refer to *PVBT* films of *ca.* 400 nm thickness with an absorbance *A* of *ca.* 0.5 at 254 nm. After irradiation with $E = 400\,\mathrm{mJ \cdot cm^{-2}}$, the amount of NCS units remained almost constant (*ca.* 31% conversion).

An estimation of the molar refraction of *PVBT* gave $R_{calc} = 50.51\,\mathrm{cm^3 \cdot mol^{-1}}$. Based on the density of *PVBT* ($\rho = 1.20\,\mathrm{g \cdot cm^{-3}}$), which was determined by a floating method at $\sim 20°$C, a refractive index of $n_D^{20} = 1.608$ is predicted for *PVBT* from the *Lorentz-Lorenz* equation. For poly-(4-vinylbenzyl-isothiocyanate), which represents a 100% conversion of SCN to NCS units, a molar refraction of $R_{calc} = 52.85\,\mathrm{cm^3 \cdot mol^{-1}}$ was computed. Assuming the density to be $1.20\,\mathrm{g \cdot cm^{-3}}$, the refractive index of this polymer would be $n_D^{20} = 1.644$. Theoretically, this refractive index would be the maximum value obtainable in *PVBT* by an UV induced isomerization of the SCN units. At 100% conversion, an increase of the

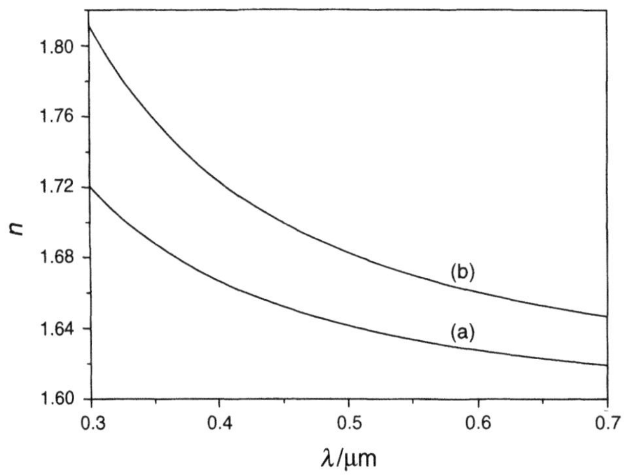

Fig. 4. Refractive index changes in a *PVBT* film; refractive index n as a function of the wavelength λ (a) before and (b) after UV irradiation

refractive index by $\Delta n_{calc} = +0.036$ would be observed. Crosslinking, which might cause an increase of the density of *PVBT*, and side-reactions, which might contribute to a change of n, are disregarded in this calculation.

Practically, we determined the refractive index of *PVBT* before and after UV irradiation by means of spectroscopic ellipsometry. For this experiment, 130 nm *PVBT* films on silicon were investigated. UV irradiation was continued until no further changes were observed in the FTIR spectrum (maximum conversion of *ca.* 31%). The dispersion of the refractive index was approximated by a *Cauchy* fit assuming an isotropic layer in each case. Figure 4 shows the refractive index for irradiated and unirradiated *PVBT* as a function of the wavelength. At $\lambda = 589$ nm we obtained $n_D^{20} = 1.629$ for unirradiated *PVBT*. After UV irradiation, the refractive index amounted to $n_D^{20} = 1.660$ ($\Delta n_{exp} = +0.031$).

The value $n_D^{20} = 1.629$, which was determined for *PVBT* by ellipsometry, is in fair agreement with the theoretical prediction. In the case of UV irradiated *PVBT*, the measured change of the refractive index ($\Delta n_{exp} = +0.031$) was lower than that calculated for 100% isomerization ($\Delta n_{calc} = +0.036$). On the one hand, the photoisomerization in *PVBT* proceeds with a yield of only *ca.* 31%; on the other hand, side reactions during illumination may contribute to changes of the refractive index.

An increase of the density of polymers during crosslinking processes is well known [18]. However, no significant increase in the density of *PVBT* was detectable after UV illumination. Summing up, the photoinduced Δn in *PVBT* is mainly a result of the partial photoisomerization of SCN to NCS units, whereas changes in the density of the polymer are of minor importance. The refractive index change measured for *PVBT* ($\Delta n_{exp} = +0.031$) is larger than that reported for other photoreactive polymers (*e.g.* poly-(vinylcinnamate): $\Delta n = -0.02$ [19]).

Stabilization of the refractive index changes

Although the refractive index changes of *PVBT* are promising for the inscription of phase gratings, the unirradiated regions of the structure are still sensitive towards

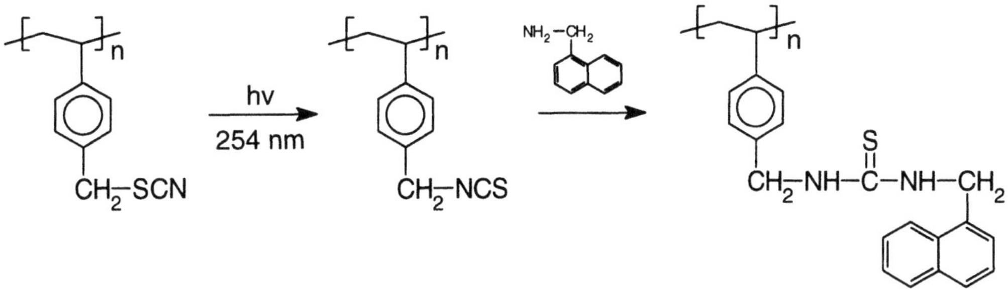

Fig. 5. Photoreaction in *PVBT* and subsequent modification with 1-naphthylmethyl amine to give a thiourea derivative

UV irradiation. This may be disadvantageous when patterned films of *PVBT* serve as a matrix material for polyconjugated materials in order to obtain lasing effects. Therefore, we searched for a way to modify the refractive index pattern so that it would remain stable even under intense UV illumination.

Isothiocyanates are known to undergo addition reactions with amines to give derivatives of thiourea. Films of *PVBT* possess the unique feature that the irradiated zones, which contain NCS groups, are acessible by amine functionalized reagents. This strategy can be employed to obtain further changes of the refractive index. On the other hand, the unirradiated zones of *PVBT* contain unchanged thiocyanate groups which show a different reactivity towards amines. This ensures a high selectivity of the derivatization process.

The modification of irradiated *PVBT* was carried out with functional groups expected to give large effects on the refractive index. Reagents containing bromine or iodine appeared unsuitable according to their low stability under UV radiation. Instead, we employed 1-aminomethyl naphthalene ($n_D^{20} = 1.642$) for the derivatization reaction shown in Fig. 5. 1-Naphthyl groups contribute to the molar refraction with $R = 43\,\mathrm{cm}^3 \cdot \mathrm{mol}^{-1}$. Thiourea units (–NH(C=S)NH–), which are formed during the addition reaction, contribute with a value of $R = 21\,\mathrm{cm}^3 \cdot \mathrm{mol}^{-1}$. If a repeating unit in *PVBT* is converted into 1-napthyl thiourea groups, a molar refraction $R_{\mathrm{calc}} = 98.5\,\mathrm{cm}^3 \cdot \mathrm{mol}^{-1}$ is calculated.

We irradiated *PVBT* films of a thickness of *ca.* 400 nm and carried out a subsequent wet chemical modification with a solution of 1-naphthylmethyl amine in methylene chloride. FTIR spectra of the modified *PVBT* film showed that the signals typical of NCS units (2085 and $2160\,\mathrm{cm}^{-1}$) decreased in intensity, whereas additional signals at $1544\,\mathrm{cm}^{-1}$ (coupled C=S and C–N vibration in thiourea) and at $794\,\mathrm{cm}^{-1}$ (*oop* deformation of C–H units in naphthalene) evolved. The ellipsometric investigation of the film revealed that the modification reaction had not occurred isotropically. The inhomogeneity of the refractive index measured at 633 nm as a function of the depth of the polymer film was approximated by a three-layer fit (Fig. 6). For this wavelength it is seen that in layers near the surface n is larger than 1.7, whereas in deeper layers the *PVBT* film retained the refractive index obtained after UV irradiation ($n \sim 1.6$).

The result is readily explained by a limited diffusion of 1-aminomethyl naphthalene into the crosslinked *PVBT* film. This gives rise to an increase of the refractive index in the layers near the surface, whereas deeper layers remain less

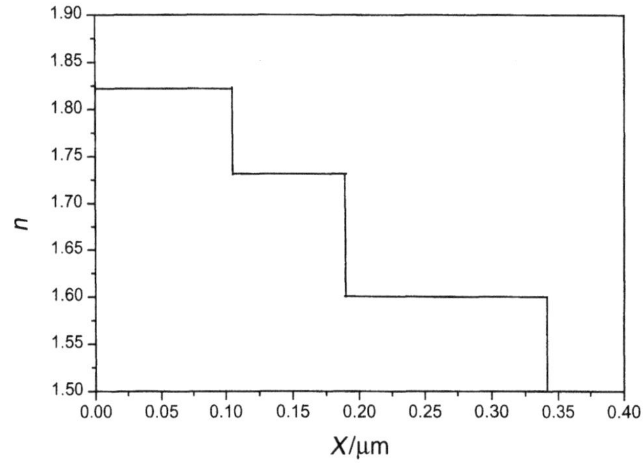

Fig. 6. Refractive index of a *PVBT* film after UV irradiation and subsequent reaction with 1-naphthylmethyl amine; three-layer fit for the refractive index n at $\lambda = 633$ nm as a function of the depth x

affected. This result demonstrates that a derivatization of UV irradiated *PVBT* with amines is useful to enhance the refractive index of the polymer film.

Conclusions

We were able to show that polymers of 4-vinylbenzyl-thiocyanate change their refractive index upon deep UV irradiation (Δn_{exp} up to +0.031). The change of the refractive index is mainly caused by a photoinduced isomerization of SCN to NCS units. The irradiated zones of *PVBT* which contain NCS groups can be modified with selected amines to give thiourea units *via* an addition reaction. Derivatization reactions of this kind are suitable to stabilize and further enhance the contrast of the refractive index in a phase-structured polymer film.

From these results we expect *PVBT* and related polymers to be suitable matrix materials for the setup of optically pumped organic lasers. In addition, these materials might be of use for deep-UV holographic recordings. Current experiments aim at lithographic fine-patterning of *PVBT* and at various modification techniques to obtain stable phase gratings which may serve as DFB resonators.

Experimental

4-Vinylbenzyl-thiocyanate (*VBT*) was synthesized according to a procedure reported earlier [26]. Poly-(4-vinylbenzyl-thiocyanate) (*PVBT*) was obtained *via* free radical polymerization of *VBT* with azo-*bis*-isobutyronitrile (*AIBN*) as initiator (*THF*, 60°C). The weight-average molecular weight M_w of *PVBT* was 220000 g · mol^{-1} as determined by size exclusion chromatography (Viscotec Model 200 differential refractometer/viscometer as detector; universal calibration). 1-Naphthylmethyl amine and benzyl isothiocyanate were bought from Fluka, benzyl thiocyanate was prepared from benzyl bromide and NH_4SCN.

The refractive indices (n_D) of liquid compounds at $\lambda = 589$ nm were measured with an *Abbe* refractometer (Krüss, Germany). Infrared spectra were taken with a Perkin-Elmer Spectrum One FTIR spectrometer. For liquid samples a KBr cuvette (100 μm; Specac) was employed. *PVBT* films were prepared by spincasting from $CHCl_3$ solutions employing a Suss RC5 spin coater. For ellipsometry, *PVBT* films (100–400 nm film thickness) were cast onto silicon wafers. The wafers were etched with HF (5%) prior to coating. Ellipsometric measurements were run with a spectroscopical ellipsometer (Sopra) [29]. The fitting procedures for ellipsometric data allowed to extract the dispersion of n in the range of 300–700 nm and the film thickness. All UV irradiations were carried out with a 1300 W high pressure Hg lamp (Heraeus) under an inert atmosphere (N_2). For monochromatic irradiation, the 254 nm line was separated with an interference filter (Melles Griot). Ferrioxalate actinometry [30] was employed to determine the intensity of the incident 254 nm light.

For the wet chemical modification of irradiated *PVBT* films the samples were immersed in a solution of 1-naphthylmethyl amine in CH_2Cl_2 (10% (w/w)). After 15 min of reaction at 20°C, the samples were extracted with CH_2Cl_2 (15 min at 20°C) and dried *in vacuo*.

Acknowledgments

Financial support by the *SFB Elektroaktive Stoffe* (Project No. 921) is gratefully acknowledged. Thanks are due to *G. Langer* and *J. Hobisch* who participated in this work as well as to *M. Winter* (*Institut für Chemische Technologie Anorganischer Stoffe*, Tu Graz) for helpful discussions.

References

[1] Burroughes JH, Bradley DDC, Brown AR, Marks RN, Mackay K, Friend RH, Burn PL, Holmes AB (1990) Nature **347**: 539; Braun D, Heeger AJ (1991) Appl Phys Lett **58**: 1982; Grem G, Leditzky G, Ullrich B, Leising G (1992) Adv Mater **4**: 36
[2] Pei Q, Gang Y, Zhang C, Yang Y, Heeger AJ (1995) Science **269**: 1086
[3] Kanicky J (1986) In: Skotheim T (ed) Handbook of Conducting Polymers. Dekker, New York, p 543
[4] Garnier F, Hajlaoui R, Yassar A, Srivastava P (1994) Science **265**: 1684
[5] Sariciftci NS, Smilowitz L, Heeger AJ, Wudl F (1992) Science **258**: 1474
[6] Hide F, Diaz-Garcia MA, Schwartz BJ, Anderson MR, Pei Q, Heeger AJ (1996) Science **273**: 1833
[7] Tessler N, Denton GJ, Friend RH (1996) Nature **382**: 695
[8] McGehee MD, Diaz-Garcia MA, Hide F, Gupta R, Miller EK, Moses D, Heeger AJ (1998) Appl Phys Lett **72**: 1536
[9] Stagira S, Zavelani-Rossi M, Nisoli M, DeSilvestri S, Lanzani G, Zenz C, Mataloni P, Leising G (1998) Appl Phys Lett **73**: 2860
[10] Berggren M, Dodabalapur A, Slusher RE, Bao Z (1997) Nature **389**: 466
[11] Kranzelbinder G, Leising G (2000) Rep Prog Phys **63**: 729
[12] Andersson MR, Yu G, Heeger AJ (1997) Synth Met **85**: 1275
[13] Frolov SV, Gellermann W, Ozaki M, Yoshino K, Vardeny ZV (1997) Phys Rev Lett **78**: 729
[14] Zenz C, Kranzelbinder G, Graupner W, Tasch S, Leising G (1999) Synth Met **101**: 222
[15] Dodabalapur A, Chandross EA, Berggren M, Slusher RE (1997) Science **277**: 1787
[16] Kogelnik H, Shank CV (1971) Appl Phys Lett **18**: 152
[17] Kallinger C, Hilmer M, Haugeneder A, Perner M, Spirkl W, Lemmer U, Feldmann J, Scherf U, Müllen K, Gombert A, Wittmer V (1998) Adv Mater **10**: 920
[18] Lessard RA, Changkakoti R, Gurusamy M (1995) Holographic Recording Materials. In: Krongauz VV, Trifunac AD (eds) Processes in Photoreactive Polymers. Chapman & Hall, New York, pp 307–367
[19] Nagata A, Sakaguchi T, Ichihashi T, Miya M, Ohta K (1997) Macromol Rapid Commun **18**: 191
[20] Biteau J, Chaput F, Lahlil K, Boilot J-P, Tsivgoulis GM, Lehn J-M, Darracq B, Marois C, Levy Y (1988) Chem Mater **10**: 1945

[21] Reiser A (1989) Photoreactive Polymers. Wiley, New York, pp 22–24
[22] Murase S, Horie K (1999) Macromolecules **32**: 1103
[23] Franke H (1992) Patterning PMMA. In: Hornak LA (ed) Polymers for Lightwave and Integrated Optics. Dekker, New York
[24] Liang J, Toussaere E, Hierle R, Levenson R, Zyss J, Ochs AV, Rousseau A, Bouitevin B (1998) Opt Mat **9**: 230
[25] Kern W, Schröder R, Hummel K, Mayer C, Hofstötter M (1998) Eur Polym J **34**: 987
[26] Kern W, Hummel K (1995) Eur Polym J **31**: 437
[27] Kern W, Hummel K (1996) Polymer **37**: 2055
[28] Kern W, Saf R, Hummel K, Kapeller H (1996) Macromol Chem Phys **197**: 329
[29] Toussaere E, Zyss J (1993) Thin Solid Films **234**: 454
[30] Hatchard CG, Parker CA (1956) Proc R Soc London **A235**: 518
[31] Parks TE, Spurlock LA (1973) J Org Chem **38**: 3922
[32] Vogel AI (1956) Practical Organic Chemistry. Longmans, London, pp 1034–1036
[33] Weast RC (ed) (1984) CRC Handbook of Chemistry and Physics. CRC Press, Boca Raton, FL

Received August 22, 2000. Accepted September 22, 2000

Mapping the Distribution of Doping Elements in Electrolytically Doped Manganese Dioxide by EFTEM and EELS

Ilse Papst[1], **Gerald Kothleitner**[1], **Ferdinand Hofer**[1,*], and **Leo Binder**[2]

[1] Forschungsinstitut für Elektronenmikroskopie und Feinstrukturforschung, Technische Universität Graz, A-8010 Graz, Austria

[1] Zentrum für Elektronenmikroskopie Graz, A-8010 Graz, Austria

[2] Institut für Chemische Technologie Anorganischer Stoffe, Technische Universität Graz, A-8010 Graz, Austria

Summary. Energy-filtering transmission electron microscopy (EFTEM) and electron energy-loss spectrometry (EELS) have been used to measure the distribution of titanium in titanium-doped electrolytic manganese dioxide, which consists of large particle agglomerates of some micrometer diameter. In contrast to previous investigations, where the distribution of the doping elements can only be investigated at the thinnest regions of the agglomerates, we describe an improved procedure which allows to measure the concentration of the dopant not only at the thinner edges of the particle agglomerate, but also in the interior. The titanium distribution maps showed that titanium is enriched on the surface of the agglomerates and in pores penetrating into their interior. Within the agglomerates, titanium is evenly distributed, and in manganese dioxide the titanium concentration could be analyzed quantitatively by EELS spectrometry. Based on these results, suggestions for possible improvements of the doping process and for the preparation of the product are given.

Keywords. Manganese dioxide; Energy-filtering transmission; Electron microscopy; Electron energy-loss spectroscopy.

Introduction

Energy-filtering transmission electron microscopy (EFTEM) has become an important new tool for the characterization of materials and biological specimen because it allows to combine morphological with chemical information [1–5]. In contrast to traditional energy-dispersive X-ray and electron energy-loss spectroscopy acquisition modes, where typically a small focused electron beam is probing the area of interest, EFTEM allows to acquire entire images that reflect for instance the two-dimensional intensity distribution of inelastically scattered electrons. These electrons give rise to an electron energy-loss spectrum (EELS), which allows the observation of electronic excitations in the target, typically within the eV to keV domain [6]. Any feature of the EELS spectrum can be used as a basis for imaging.

* Corresponding author

Perhaps the most frequently used application technique in EFTEM is to select ionization edges in order to form elemental maps. One of the main advantages of this technique is that elemental maps can be recorded with high spatial resolution in the nanometer range, rendering it particularly well suited to explore local chemical and electronic properties of nanocrystalline and nanostructured materials at dimensions in the nanometer regime [7].

In a previous investigation, EFTEM has successfully been used to measure the distribution of titanium in titanium-doped electrolytic manganese dioxide (EMD) [8]. Adding dopants to EMD is important for stabilizing the structure of MnO_2 as used in secondary batteries as the major component of composite cathodes in rechargeable alkaline manganese dioxide (RAM^{TM}) cells [9]. However, the rechargeability of these batteries depends on a partial discharge of the manganese dioxide cathode. Therefore, many attempts have been made to improve the possible depth-of-discharge and the number of charge/discharge cycles. Evidently, there is a clear correlation between structural parameters of manganese dioxide and its rechargeability, which can be improved by physically or chemically modifying the MnO_2 material. A successful method for stabilizing the structure of EMD is to dope it with Ti(IV) cations. Initial work on titanium doping effects utilised tetrachloro-titanium ($TiCl_4$) as a doping agent. This compound is very reactive, hydrolyzing even in the presence of atmospheric water vapour, and therefore had to be handled under a protective layer of nitrogen gas. Other tested doping materials were tetra-*n*-propoxytitanium ($Ti(OC_3H_7)_4$), tetra-*n*-butoxytitanium ($Ti(OC_4H_9)_4$), and titanium oxysulfate ($TiO(SO_4)$), most of them being highly unstable. Even the organotitanium compounds slowly hydrolyze in aqueous media, and titanium oxysulfate is only stable at higher concentrations of sulfuric acid. This poses a problem because under these conditions the side reaction of permanganate formation instead of manganese dioxide deposition becomes dominant, a fact which is quite cumbersome and not very convenient for Ti-EMD production on an industrial scale. Therefore, we have put the focus of this work on the investigation of dopants introduced *via* more simple precursors that are easier to handle in industrial manufacturing. Overall, the effect of introducing Ti(IV) cations could increase the number of possible cycles by over 100% compared with standard EMDs [10].

Consequently, for optimizing the doping process it is necessary to know the distribution of the doping element in the starting EMD material and in cycled EMDs. As already mentioned, EFTEM can in principle provide this kind of information. However, what complicates the situation for EFTEM investigations is the fact that EMD nanocrystalline particles form rather large agglomerates of some micrometers in diameter that are often too thick to be suitable for EFTEM studies.

In this paper we report a new preparation technique for thin TEM samples that helps to avoid the thickness limitations of EFTEM in order to get detailed quantitative information on the distribution of dopants in EMD.

Results and Discussion

A previously carried out X-ray diffraction experiment did not reveal significant differences between doped and undoped EMD samples, except that there was a

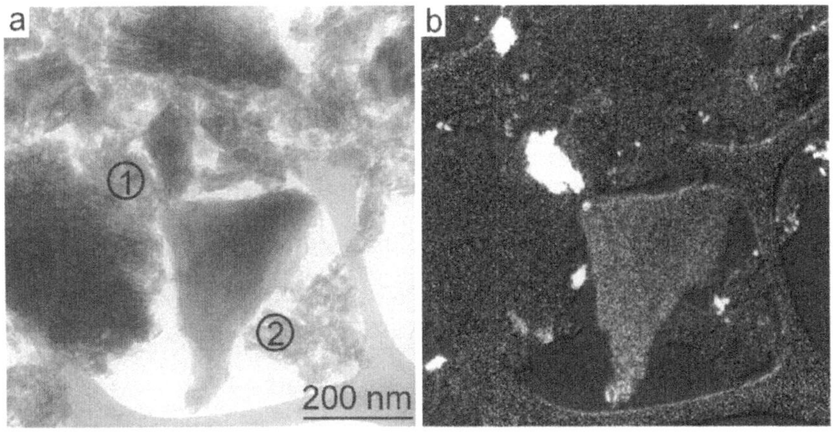

Fig. 1. (a) TEM image of titanium-doped EMD (laboratory product) with areas of EELS analyses indicated; (b) titanium distribution map recorded with the Ti-$L_{2,3}$ ionization edge of the same specimen region as shown in Fig. 1a

slight shift of some X-ray lines relative to those of the manganese dioxide [8]. This shift could be attributed to a distortion of the crystal lattice, but was not followed up in further detail. Atomic absorption spectroscopy (AAS) was used to measure the titanium concentration in EMD samples, and organotitanium doped EMD was found to take up an average concentration of the doping element of about 1% (w/w).

However, X-ray diffraction and atomic absorption spectroscopy can neither reveal whether all the titanium is incorporated into the manganese dioxide, or if it is mainly enriched in Ti-rich phases, nor how the chemical composition is influenced during cycling. Therefore, an analytical transmission electron microscope (AEM) is an ideal tool for answering these type of questions with respect to titanium doped EMD.

A typical TEM image (Fig. 1a) of the uncycled laboratory product (doped with $Ti(OBu)_4$) shows that the powder is composed of particles with an average diameter of about 200 nm. These relatively large powder particles consist of nanosized manganese dioxide crystals. The corresponding titanium distribution map (jump ratio image (Fig. 1b)) indicates some agglomerations of titanium-rich material (bright regions), whereas most of the incorporated titanium is homogenously distributed.

In order to find out more about the chemical composition of the various phases, EELS spectra were recorded from various specimen regions as marked in Fig. 1a. These spectra (Fig. 2) can also be used to derive quantitative composition information. In the spectrum from the titanium-rich bright particle (region 1), edges from Ti ($L_{2,3}$), O (*K*), and Mn ($L_{2,3}$) occur. This particular phase corresponds to a Ti-Mn-oxide with an Ti/O atomic ratio of 0.41 and an Mn/O atomic ratio of 0.27 (Ti/Mn = 2.56). The spectrum recorded at the specimen region 2 originates from the MnO_2 matrix and exhibits only a small Ti-$L_{2,3}$ edge which yields a Ti/Mn atomic ratio of about 0.004. Segregations of pure titanium dioxide could not be found. The

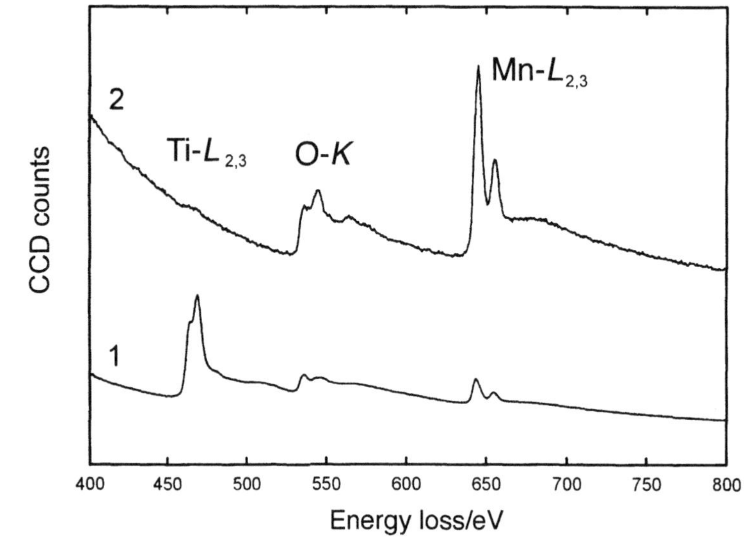

Fig. 2. EELS spectra recorded at the specimen regions indicated in Fig. 1a

main conclusion from this work is that Ti is not completely incorporated into manganese dioxide when organotitanium dopants are used.

However, as already mentioned, due to thickness limitations, conventional EFTEM/EELS investigations on powder samples allow to analyze the chemistry of the thinnest parts only of the rather big particle agglomerates, which are some tenths of a micrometer in size. It is not possible to verify a homogenous Ti distribution in the interior of the agglomerates. In particles not much thicker than 0.1 μm, multiple electron scattering already degrades the element-specific signal, but in specimens of more than 0.3 μm thickness not even 200 kV electrons can successfully traverse the sample and hence get absorbed (black particles in Fig. 1a).

Embedding the powder sample in epoxy resin with subsequent ion-milling is a viable way to investigate both the surface and the interior of particle agglomerates. Figure 3 represents such a TEM cross-section through a big particle agglomerate of the TiO_2-doped material, which again consists of small nanometer-sized crystallites. The TEM bright field image in Fig. 4a shows a typical specimen region of the same sample. The absolute thickness map (Fig. 4b) clearly reveals that the sample thickness of the cross-sectioned particles falls within a range of 35 and 85 nm and hence does not change significantly in the field of view. Because multiple scattering is not a predominant process in this thickness range, EELS and EFTEM can be reliably applied to study elemental distributions within the agglomerates.

By comparing the two elemental maps for titanium and manganese from Figs. 4c and 4d (Ti-$L_{2,3}$ and Mn-$L_{2,3}$ maps) it becomes clear that there are Ti-richer areas on the surface of the EMD agglomerates. These regions appear bright in the titanium map and consist of Ti-Mn-oxides. However, the titanium map also shows a homogeneous distribution of titanium in the interior of the agglomerates. The titanium concentration at the interior of various agglomerates was measured by EELS spectroscopy. A quantitative EELS analysis of a spectrum taken at the

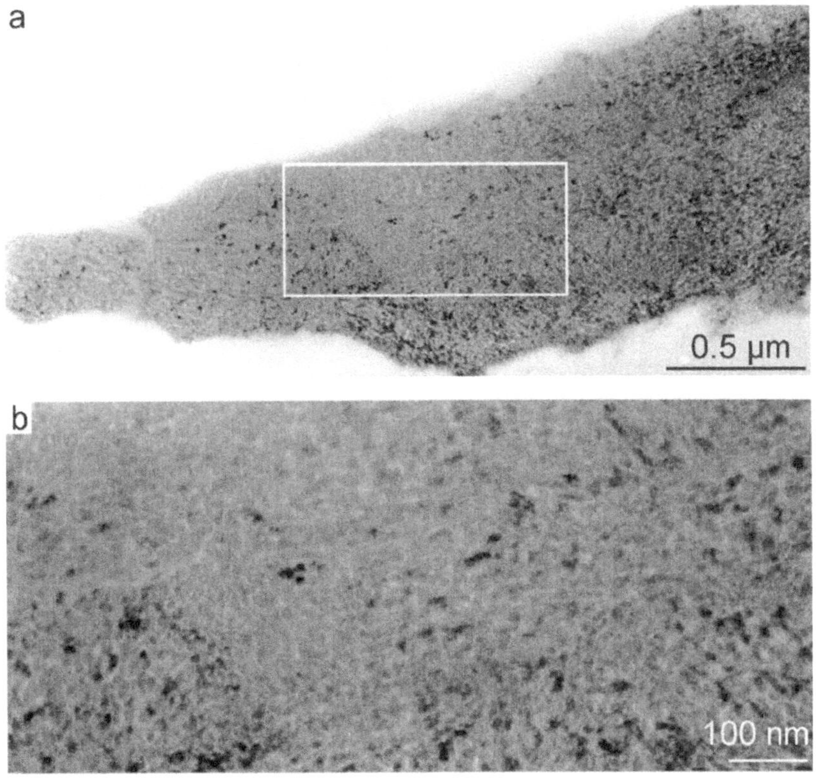

Fig. 3. TEM image of a cross-sectioned EMD agglomerate revealing the internal structure

interior of these agglomerates (marked region in Fig. 4a; Fig. 5) gave a Ti/Mn atomic ratio of about 0.05. This ratio is much higher than that determined for the organotitanium doped EMD [8]. There is clear evidence for titanium enrichment at the surface of the agglomerates as well as in the pores that penetrate into the agglomerates (indicated by the arrows in Fig. 4c). A concentration gradient of the dopant within the agglomerates could not be found.

The Ti/Mn ratios of the inner part of the EMD agglomerates have been measured for $TiOSO_4$- and TiO_2-doped EMD, both in the uncycled and cycled state, and are summarized in Table 1.

These experimental findings suggest that cycling the material increases the Ti concentration in the large EMD agglomerates. Furthermore, the TEM images indicate that the size and morphology of the nanometer-sized grains in the EMD was not considerably changed by cycling. In addition, EMDs doped with $TiOSO_4$ contain a significantly higher titanium concentration than those doped with TiO_2. Comparison of the EFTEM elemental maps indicated an equally even distribution of titanium in both types of EMD samples ($TiOSO_4$- and TiO_2-doped). A repeated charging and discharging of both types of EMD as a cathode material in a rechargeable manganese dioxide/zinc-cell (cycling about 40 times) led to practically the same titanium distribution. The reason for this effect is the low solubility of titanium dioxide in the concentrated KOH solution ($9\,M$) serving as the battery electrolyte.

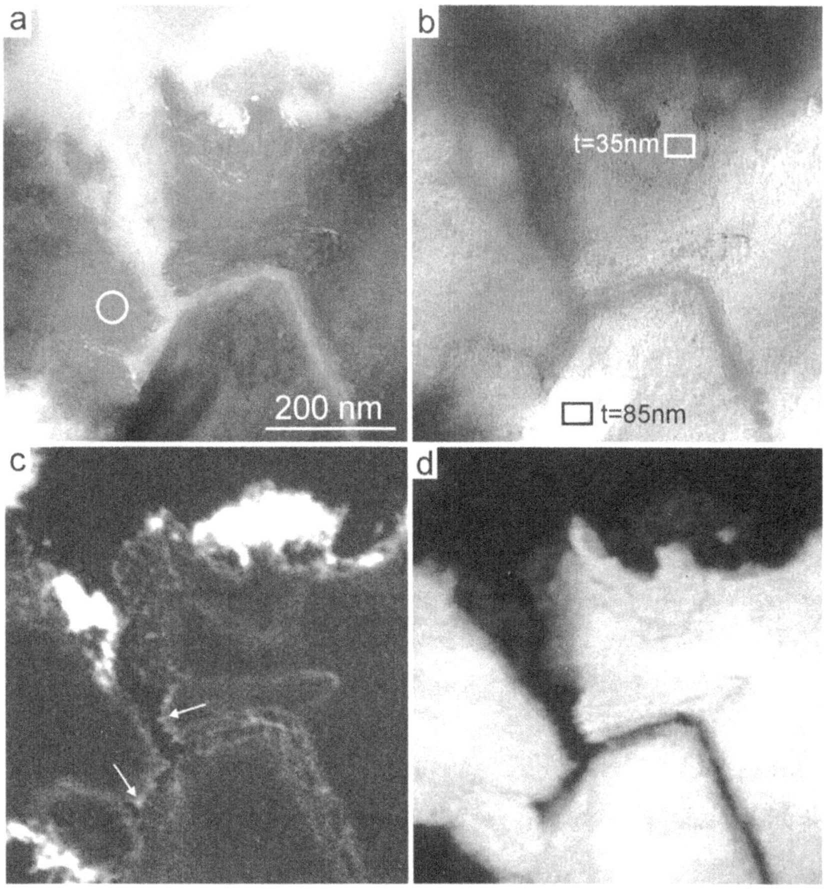

Fig. 4. (a) TEM image of a cycled titanium doped EMD prepared by cross-sectioning with areas of EELS analyses indicated; (b) EFTEM thickness map (t/λ); (c) titanium distribution map recorded with the Ti-$L_{2,3}$ ionization edge; (d) manganese distribution map recorded with the Mn-$L_{2,3}$ ionization edge

Equal amounts of titanium dopant in both types of EMDs clearly caused more Ti-Mn-oxide particles to be included in the sample made from titanium dioxide than in EMDs produced by doping with $TiOSO_4$. Thus, titanium mapping indicated accumulation of Ti at particular spots on the surface of the agglomerates of the TiO_2-doped sample. Future experiments will shed some light on the question of whether a smaller fraction of the doping compound in the electrolyte during EMD deposition may result in a truly homogeneous distribution of titanium in the product.

Sample preparation by means of ion-milling bigger agglomerates, which are normally too thick for EFTEM studies, enables the preparation of useable electron transparent cross-sections. This preparation technique opens up a simple but efficient route for studying the chemistry of sub-μm powders such as catalyst supports or precursor materials, which are not directly accessible by surface analysis techniques or microprobe analysis.

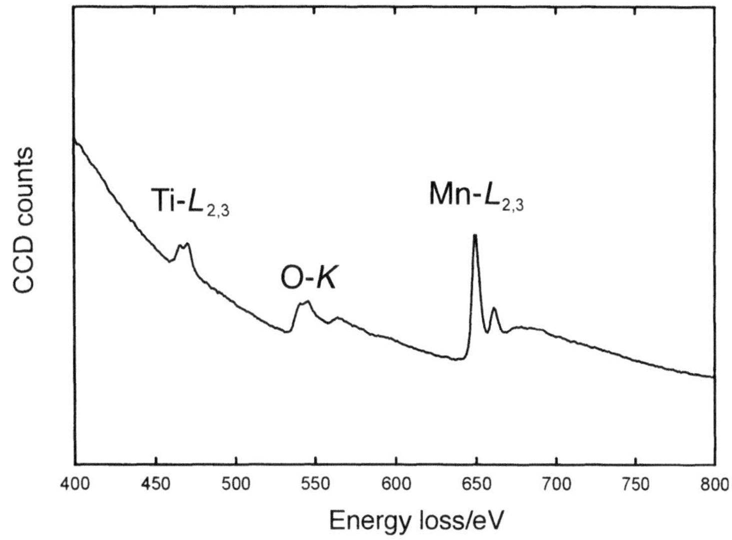

Fig. 5. EELS spectrum recorded at the specimen region indicated in Fig. 4a

Table 1. Ti/Mn atomic ratios of the interior of the large EMD agglomerates measured by EEL spectrometry

Ti/Mn	Uncycled	Cycled
$TiOSO_4$-doped EMD	0.07 ± 0.01	0.22 ± 0.02
TiO_2-doped EMD	0.015 ± 0.003	0.05 ± 0.01

Conclusions

EFTEM elemental mapping is a powerful tool for studying the distribution and concentration of dopants in electrolytically doped manganese dioxides. Ion-milling of big EMD agglomerates, which are too thick for electron penetration, enables the visualization of the distribution of the dopants from the surface to the interior and the measurement of the dopant concentration on the surface and in the interior. This information cannot be obtained from conventionally prepared TEM specimens.

All titanium doped EMDs were found to contain an excess of the doping element, which was enriched in the form of Ti-Mn-oxides either on the surface or in the pores of the agglomerates. However, the major fraction of the titanium was evenly distributed in the interior of the agglomerates, and the dopant concentration clearly depends on the doping route used for the starting material. It was found that the dopant concentration can be increased by cycling of the EMD.

Experimental

Production of test samples

For the production of the samples, a laboratory scale electrolyzer was used. A four-necked $4\,dm^3$ round bottom flask equipped with inlets for a reflux condenser, a thermometer, and the electrodes served as the electrolytic vessel. The electrode arrangement consisted of a three-electrode system made up of one anode and two cathodes which were separated by spacers made of *PVC*. The anode was made up of a $9 \times 9.5\,cm^2$ titanium metal sheet; the cathodes (same dimension) were made from lead metal sheets.

For the preparation of the EMD, a $3\,dm^3$ volume of the basic electrolyte solution mixed with $50\,cm^3/dm^3$ of (liquid) dopant was introduced into the electrolyzer. The basic electrolyte solution consisted of $112\,g/dm^3$ $MnSO_4 \cdot H_2O$ and $98\,g/dm^3$ H_2SO_4.

Titanium doped electrolytic manganese dioxide (Ti-EMD) samples were prepared by means of a suitable organotitanium compound (tetrabutyltitanate, $Ti(OBu)_4$, as liquid dopant), $TiOSO_4$, and fine-grained titanium dioxide ($0.5\,g/dm^3$ solid dopant applied in suspension) as additives to the electrolyte.

For all materials, the anode current density was maintained constant at about $6\,mA/cm^2$ for a 48 h period of electrolysis. During electrolysis, the temperature was maintained at 98°C in order to permit the predominant formation of γ-EMD.

The electrolysis bath was recharged after 24 h by introducing 40 g of $MnCO_3$ stepwise and in small quantities. This technique replenishes the electrolyte with an approximate quantity of manganese removed by electrolysis and provides neutralization of formed protons. The replenishment step is necessary to avoid electrochemical decomposition of EMD deposited at the anode.

At the end of the 48 h period of electrolysis, the electrode system was removed and disassembled, and the anode was removed and washed thoroughly with distilled water. The deposited Ti-EMD was mechanically removed and ground with a pestle and mortar. The Ti-EMD powder obtained was washed well with $1\,mol/dm^3$ KOH followed by sufficient distilled water until the washing solution reacted nearly neutral (*pH* 6.5). The Ti-EMD was recovered by filtering under suction and dried at 80°C overnight. The product was sieved using 125 μm mesh screen and stored for characterization (detailed description in Ref. [11]).

TEM specimen preparation and TEM investigation

For TEM sample preparation two different techniques were applied. In a first approach the uncycled Ti-doped EMD powder specimen (doped with organotitanium compounds) was mounted on holey carbon grids following standard TEM preparation procedures. In a second approach, powder of the TiO_2-cycled material was embedded in epoxy resin, mechanically grinded and dimpled, and finally argon milled under low angle [12].

The instrument used for TEM investigations was a Philips CM20/STEM operated with a LaB_6 cathode at an acceleration voltage of 200 kV. The microscope was equipped with a post-column imaging filter (GIF by Gatan) which allows both the acquisition of electron energy-loss spectra (EELS) and energy-filtered images (EFTEM). All images and spectra were recorded with the slow-scan CCD camera part of the GIF and processed with Gatan's DigitalMicrograph and EL/P software.

Elemental mapping was then carried out *via* scripts implemented in DigitalMicrograph. To obtain an elemental map, two energy filtered background images in front of the edge were acquired together with one image at the ionization edge of the element of interest. Then, an extrapolated background image was calculated using the power law model $I = AE^{-r}$, where I is the intensity, E the energy loss, and A and r are two fitting parameters. This background image is subtracted from the ionization edge image, giving net intensities that relate to elemental concentration. Three-window elemental mapping, however, has some shortcomings [13, 14], and therefore an alternative method was developed to obtain elemental distribution images. By simply dividing the ionization edge image with a pre-edge image (so called jump ratio images) one can get a similar type of information [15].

An important parameter, which significantly determines the detectability of elements in EFTEM, is the specimen thickness, determining the degree of plural scattering of transmitted electrons and hence affecting the visibility of edges (signal-to-background ratio) in EELS spectra. In order to monitor the thickness of specimens and to better study the quality of the sample preparation, we made use of the log ratio technique during our experiments. This technique is based on the fact that the probability (*i.e.* spectral intensity) of an electron to undergo multiple inelastic scattering events in a specimen of a certain thickness and a given inelastic mean free path λ can be described by *Poisson* statistics. As a consequence of this behaviour, a logarithmic image, calculated by forming the ratio of two other images in which one contains all elastically scattered electrons and the other one the total spectral intensity, yields a representation of the sample's thickness relative to its inelastic mean free path [16]. The λ value for MnO_2 can be estimated from the formula given by *Malis* and co-workers [17] and amounts to approximately 125 nm for a collection angle of 7.6 mrad and a primary energy of 200 kV as used in our experimental setup. By multiplying the pixel intensities with the inelastic mean free path, one can easily convert relative thickness values into absolute sample thicknesses.

Acknowledgements

The authors gratefully acknowledge financial support from the *Fonds der wissenschaftlichen Forschung* (FWF) within the Special Research Program *Electroactive Materials*. The authors thank *Albert Brunegger* and *Peter Warbichler* for the preparation of the TEM specimens.

Grateful acknowledgement is made to the companies Rhodia, TiO_2 Specialities Branch, Aubervilliers, France, and Millenium, Inorganic Chemicals, Saint-Denis, France, for the supply with titanium chemicals.

References

[1] Hofer F, Grogger W, Warbichler P, Papst I (2000) Microchimica Acta **132**: 273
[2] Leapman RD, Hunt JA (1995) J Microsc Soc Am **1**: 93
[3] Grogger W, Hofer F, Warbichler P, Kothleitner G (2000) Microsc Microanal **6**: 161
[4] Beckers ALD, De Bruijn WC, Cleton-Soeteman MI, Van Eijk HG, Gelsema ES (1997) Micron **28**: 349
[5] Mayer J, Eigenthaler U, Plitzko JM, Dettenwanger F (1997) Micron **28**: 361
[6] Egerton RF (1996) Electron-Energy Loss Spectroscopy in the Electron Microscope. Plenum, New York
[7] Reimer L (1995) Energy-Filtering Transmission Electron Microscopy. Springer, Berlin
[8] Binder L, Jantscher W, Hofer F, Kothleitner G (1998) J Power Sources **70**: 1
[9] Binder L, Odar W, Kordesch K (1981) J Power Sources **6**: 271
[10] Urdl P (1994) PhD Thesis, University of Technology, Graz, Austria
[11] Nartey VK, Binder L, Huber A (2000) J Power Sources **87**: 205
[12] Aldrian A, Brunegger A, Warbichler P, Windisch G (1996) Proc 11th Eur Congr on Electron Microscopy, Dublin, p 746
[13] Krivanek OL, Gubbens AJ, Delby N, Meyer CE (1993) Microsc Microanal Microstruct **3**: 187
[14] Hofer F, Warbichler P, Grogger W (1995) Ultramicroscopy **59**: 15
[15] Johnson DE (1979) Introduction to Analytical Electron Microscopy. Plenum Press, New York, p 245
[16] Hofer F, Grogger W, Kothleitner G, Warbichler P (1997) Ultramicroscopy **67**: 83
[17] Malis T, Cheng SC, Egerton RF (1988) J Electr Microsc Techn **8**: 193

Received November 29, 2000. Accepted (revised) December 12, 2000

GPSR Compliance

The European Union's (EU) General Product Safety Regulation (GPSR) is a set of rules that requires consumer products to be safe and our obligations to ensure this.

If you have any concerns about our products, you can contact us on ProductSafety@springernature.com

In case Publisher is established outside the EU, the EU authorized representative is:

Springer Nature Customer Service Center GmbH
Europaplatz 3
69115 Heidelberg, Germany

Batch number: 10585278

Printed by Printforce, the Netherlands